A HANDBOOK OF
LABORATORY GLASS-BLOWING

To my Friends
Eric Reid
and
Sidney Wilkinson

A Handbook of Laboratory Glass-Blowing

BY

BERNARD D. BOLAS

WITH NUMEROUS DIAGRAMS IN THE TEXT
BY NAOMI BOLAS

LONDON
GEORGE ROUTLEDGE & SONS, LTD.
New York: E. P. DUTTON & CO.
1921

CONTENTS

CHAP.   PAGE

I. Introduction and Preliminary Remarks — General Principles to be observed in Glass Working— Choice of Apparatus—Tools and Appliances— Glass   1

II. Easy Examples of Laboratory Glass-Blowing — Cutting and Sealing Tubes, Tubes for High Temperature Experiments—Thermometer-Bulbs Bulbs of Special Glass, Pipettes, Absorption-Bulbs or Washing Bulbs — Joining Tubes, Branches, Exhaustion-Branches, Branches of Dissimilar Glass, Blowing Bulbs, A Thistle Funnel, Cracking and Breaking Glass, Leading and Direction of Cracks— Use of Glass Rod or Strips of Window-Glass, Joining Rod, Feet and Supports—Gripping Devices for use in Corrosive Solutions—The Building up of Special Forms from Solid Glass   10

III. Internal Seals, Air-Traps, Spray Arresters, Filter-Pumps—Sprays, Condensers ; plain, double surface, and spherical—Soxhlet Tubes and Fat Extraction Apparatus—Vacuum Tubes, Electrode Work, Enclosed Thermometers, Alarm Thermometers . . . Recording Thermometers, "Spinning" Glass   32

IV. Glass, its Composition and Characteristics—Annealing—Drilling, Grinding, and Shaping Glass by methods other than Fusion — Stop-cocks —

CONTENTS

CHAP.		PAGE
	Marking Glass—Calibration and Graduation of Apparatus—Thermometers—Exhaustion of Apparatus—Joining Glass and Metal—Silvering Glass	55
V.	Extemporised Glass-Blowing Apparatus—The use of Oil or other Fuels—Making Small Rods and Tubes from Glass Scraps—The Examination of Manufactured Apparatus with a view to Discovering the Methods used in Manufacture—Summary of Conditions necessary for Successful Glass-Blowing	80
Index		105

PREFACE

To cover the whole field of glass-blowing in a small handbook would be impossible. To attempt even a complete outline of the methods used in making commercial apparatus would involve more than could be undertaken without omitting the essential details of manipulation that a novice needs. I have, therefore, confined myself as far as possible to such work as will find practical application in the laboratory and will, I hope, prove of value to those whose interests lie therein.

The method of treatment and somewhat disjointed style of writing have been chosen solely with the view to economy of space without the undue sacrifice of clearness.

BERNARD D. BOLAS.

Handbook of Laboratory Glass-Blowing

CHAPTER I

Introduction and Preliminary Remarks—General Principles to be observed in Glass Working—Choice of Apparatus—Tools and Appliances—Glass.

GLASS-BLOWING is neither very easy nor very difficult; there are operations so easy that the youngest laboratory boy should be able to repeat them successfully after once having been shown the way, there are operations so difficult that years are needed to train eye and hand and judgment to carry them out; but the greater number of scientific needs lie between these two extremes. Yet a surprisingly large number of scientific workers fail even to join a glass tube or make a T piece that will not crack spontaneously, and the fault is rather one of understanding than of lack of ability to carry out the necessary manipulation.

In following the scheme of instruction adopted in this handbook, it will be well for the student to pay particular attention to the reason given for each detail of the desirable procedure, and, as far as may be, to memorise it. Once having mastered the underlying reason, he can evolve schemes of manipulation to suit his own particular needs, although, as a rule, those given in the following pages will be found to embody the result of many years' experience.

There is a wide choice of apparatus, from a simple mouth-blowpipe and a candle flame to a power-driven blower and a multiple-jet heating device. All are useful, and all have their special applications, but, for the present, we will consider the ordinary types of bellows and blowpipes, such as one usually finds in a chemical or physical laboratory.

The usual, or Herepath, type of gas blow-pipe consists of an outer tube through which coal gas can be passed and an inner tube through which a stream of air may be blown. Such a blowpipe is shown in section by Fig. 1. It is desirable to have the three centring screws as shown, in order to adjust the position of the air jet and obtain a well-shaped flame, but these screws are sometimes omitted. Fig. 1, *a* and *b* show the effects of defective centring of the air jet, *c* shows the effect of dirt or

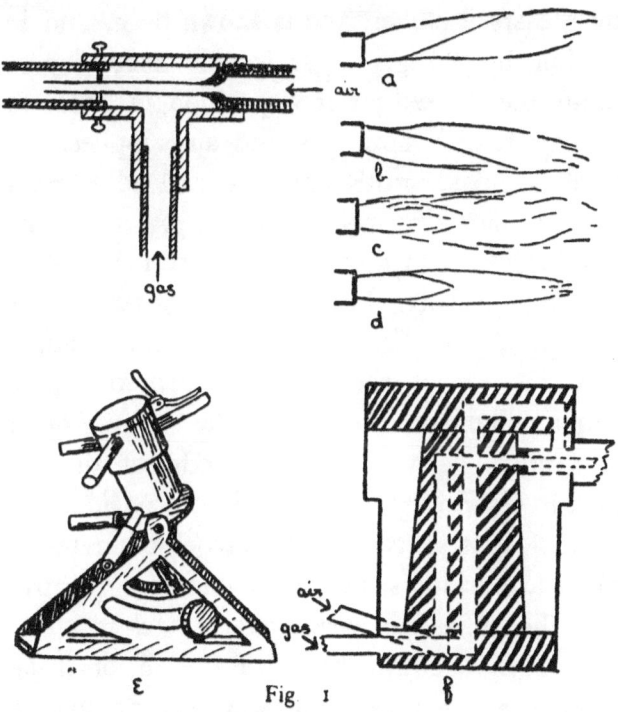

Fig 1

roughness in the inside of the air jet, *d* shows a satisfactory flame.

For many purposes, it is an advantage to

have what is sometimes known as a "quick-change" blowpipe; that is one in which jets of varying size may be brought into position without stopping the work for more than a fraction of a second. Such a device is made by Messrs. Letcher, and is shown by *e*, and in section by *f* Fig. 1. It is only necessary to rotate the desired jet into position in order to connect it with both gas and air supplies. A small bye-pass ignites the gas, and adjustment of gas and air may be made by a partial rotation of the cylinder which carries the jets.

For specially heavy work, where it is needed to heat a large mass of glass, a multiple blowpipe jet of the pattern invented by my father, Thomas Bolas, as the result of a suggestion derived from a study of the jet used in Griffin's gas furnace, is of considerable value. This jet consists of a block of metal in which are drilled seven holes, one being central and the other six arranged in a close circle around the central hole. To each of these holes is a communication way leading to the gas supply, and an air jet is arranged centrally in each. Each hole has also an extension tube fitted into it, the whole effect being that of seven blowpipes. In order to provide a final adjustment for the flame, a perforated plate having seven holes which correspond in size and position to the outer tubes is arranged to slide on parallel guides in front of these outer tubes.

The next piece of apparatus for consideration is the bellows, of which there are three or more types on the market, although all consist of two essential parts, the blower or bellows proper and the wind chamber or reservoir. Two patterns are shown in Fig. 2; *a*, is the

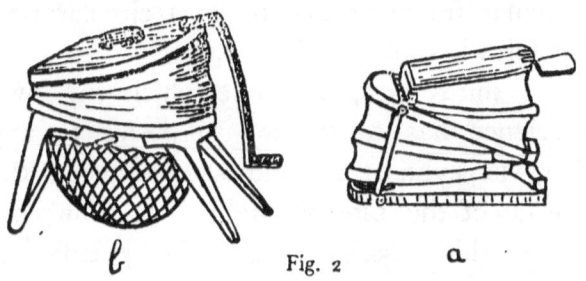

Fig. 2

form which is commonly used by jewellers and metal workers to supply the air blast necessary for heating small furnaces. Such a bellows may be obtained at almost any jewellers' supply dealer in Clerkenwell, but it not infrequently happens that the spring in the wind chamber is too strong for glass-blowing, and hence the air supply tends to vary in pressure. This can be improved by fitting a weaker spring, but an easier way and one that usually gives fairly satisfactory results, is to place an ordinary screw-clip on the rubber tube leading from the bellows to the blowpipe, and to tighten this until an even blast is obtained.

Another form of bellows, made by Messrs. Fletcher and Co., and common in most laboratories, is shown by b; the wind chamber consists of a disc of india-rubber clamped under a circular frame or tied on to a circular rim. This form is shown by Fig. 2, b.

The third form, and one which my own experience has caused me to prefer to any other, is cylindrical, and stands inside the pedestal of the blowpipe-table. A blowpipe-table of this description is made by Enfer of Paris.

There is no need, however, to purchase an expensive table for laboratory use. All the work described in this book can quite well be done with a simple foot bellows and a quick-change blowpipe. Nearly all of it can be done with a single jet blowpipe, such as that described first, or even with the still simpler apparatus mentioned on page 84, but I do not advise the beginner to practise with quite so simple a form at first, and for that reason have postponed a description of it until the last chapter.

Glass-blowers' tools and appliances are many and various, quite a number of them are better rejected than used, but there are a few essentials. These are,—file, glass-knife, small turn-pin, large turn-pin, carbon cones, carbon plate, rubber tube of small diameter, various sizes of corks, and an asbestos heat reflector. For ordinary work, an annealing oven is not necessary, but one is described on page 60 in connection with the special cases where annealing is desirable.

Fig. 3 illustrates the tools and appliances. a is an end view of the desirable form of file, and shows the best method of grinding the edges in order to obtain a highly satisfactory tool. b is a glass knife, shown both in perspective and end view, it is made of glass-hard steel and should be sharpened on a rough stone, such as a scythe-stone, in order to give a slightly irregular edge. c is a small turn-pin which may be made by flattening and filing the end of a six-inch nail. *d* is the large turn-pin and consists of a polished iron spike, about five inches long and a quarter of an inch diameter at its largest part. This should be mounted in a wooden handle. *e* and *f* are carbon cones. A thin rubber tube is also useful; it may be at-

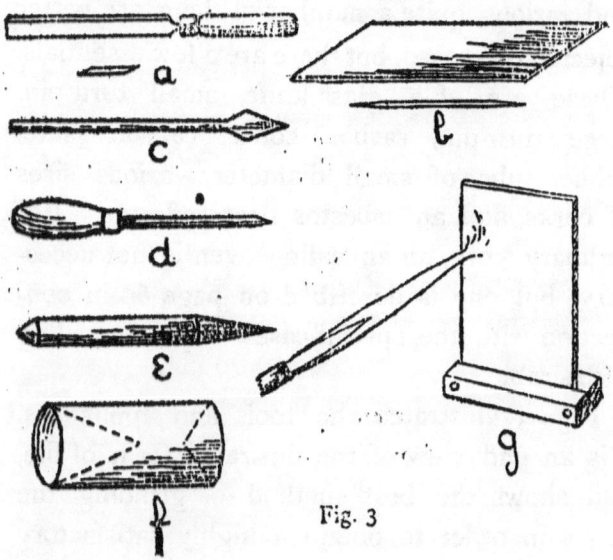

Fig. 3

tached to the work and serve as a blowing tube, thus obviating the necessity of moving the work to the mouth when internal air pressure is to be applied. In order to avoid undue repetition, the uses of these tools and appliances will be described as they occur.

Glass, as usually supplied by chemical apparatus dealers is of the composition known as "soda-glass." They also supply "hard" or "combustion" glass, but this is only used for special purposes, as it is too infusible for convenient working in the ordinary blowpipe flame.

Soda-glass consists primarily of silicate of sodium with smaller quantities of silicate of aluminum and potassium. Its exact composition varies. It is not blackened, as lead glass is, by exposure to the reducing gases which are present in the blue cone of a blowpipe flame, and hence is easier for a beginner to work without producing discolouration.

Further notes on glasses will be found on page 55, but for ordinary purposes soda-glass will probably be used.

CHAPTER II

Easy Examples of Laboratory Glass-Blowing—Cutting and Sealing Tubes for Various Purposes; Test-Tubes, Pressure-Tubes, Tubes for High Temperature Experiments—Thermometer-Bulbs, Bulbs of Special Glass, Pipettes, Absorption-Bulbs or Washing-Bulbs—Joining Tubes; Branches, Exhaustion-Branches, Branches of Dissimilar Glass—Blowing Bulbs; A Thistle Funnel; Cracking and Breaking Glass; Leading and Direction of Cracks—Use of Glass Rod or Strips of Window-Glass; Joining Rod, Feet and Supports—Gripping Devices for use in Corrosive Solutions—The Building Up of Special Forms from Solid Glass.

PERHAPS the most common need of the glass-blower whose work is connected with that of the laboratory is for a sealed tube; and the sealing of a tube is an excellent preliminary exercise in glass-blowing.

We will assume that the student has adjusted the blowpipe to give a flame similar to that shown in *d*, Fig. 1, and that he has learned to maintain a steady blast of air with the bellows; further, we will assume that the tube he wishes to seal is of moderate size, say not more than half an inch in diameter and with walls of from one-tenth to one-fifth of an inch thick.

A convenient length of tube for the first trial is about one foot; this should be cut off from the longer piece, in which it is usually supplied, as follows:—lay the tube on a flat surface and make a deep cut with the edge of a file. Do

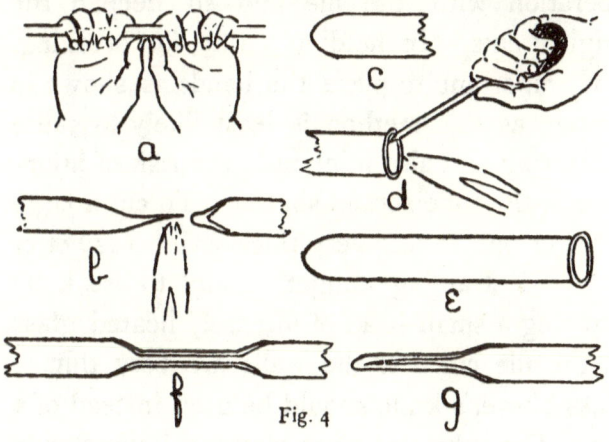

Fig. 4

not "saw" the file to and fro over the glass. If the file edge has been ground as shown in *a*, Fig. 3, such a procedure will be quite unnecessary and only involve undue wear; one movement with sufficient pressure to make the file "bite" will give a deep cut. Now rotate the tube through about one-eighth of a turn and make another cut in continuation of the first. Take the tube in the hands, as shown in *a*, Fig. 4, and apply pressure with the thumbs, at the same time straining at the ends. The tube should break easily. If it does not, do not strain too hard, as it may shatter and cause serious injuries to the hands, but repeat the operation with the file and so deepen the original cuts. In holding a tube for breaking, it is important to place the hands as shown in sketch, as this method is least likely to cause shattering and also minimises the risk of injury even if the tube should shatter. To cut a large tube, or one having very thick walls, it is better to avoid straining altogether and to break by applying a small bead of intensely heated glass to the file cut. If the walls are very thin, a glass-blower's knife should be used instead of a file. The tube and glass-blower's knife should be held in the hand, and the tube rotated against the edge of the knife; this will not produce a deep cut, but is less likely to break the tube. A bead of hot glass should be used to complete the work.

The next operation is to heat the glass tube in the middle; this must be done gradually

LABORATORY GLASS-BLOWING

and evenly; that is to say the tube must be rotated during heating and held some considerable distance in front of the flame at first; otherwise the outer surface of the glass will expand before the interior is affected and the tube will break. From two to five minutes, heating at a distance of about eight inches in front of the flame will be found sufficient in most cases, and another minute should be taken in bringing the tube into the flame. Gradual heating is important, but even heating is still more important and this can only be obtained by uniform and steady rotation. Until the student can rotate a tube steadily *without thinking about it*, real progress in glass-blowing is impossible.

When the tube is in the flame it must be held just in front of the blue cone and rotated until the glass is soft enough to permit the ends to be drawn apart. Continue to separate the ends and, at the same time, move the tube very slightly along its own axis, so that the flame tends to play a little more on the thicker part than on the drawn-out portion. If this is done carefully, the drawn-out portion can be separated off, leaving only a slight "bleb" on the portion it is desired to seal. This is illustrated by *b*, Fig. 4.

To convert the seal at *b*, Fig. 4., into the ordinary form of test-tube seal, it is only necessary to heat the "bleb" a little more strongly, blow gently into the tube until the thick portion is slightly expanded, re-heat the whole of the rounded end until it is beginning to collapse, and give a final shaping by careful blowing after it has commenced to cool. In each case the glass must be removed from the flame before blowing. The finished seal is shown by *c*, Fig. 4. If desired, the open end may now be finished by heating and rotating the soft glass against the large turn-pin, as illustrated in *d*, but the turn-pin must not be allowed to become too hot, as if this happens it will stick to the glass. After turning out the end, the lip of glass must be heated to redness and allowed to cool without coming in contact with anything; otherwise it will be in a condition of strain and liable to crack spontaneously. The finished test-tube is shown by *e*.

When it is necessary to seal a substance inside a glass tube, the bottom of the tube is

LABORATORY GLASS-BLOWING

first closed, as explained above, and allowed to cool; the substance, if a solid, is now introduced, but should not come to within less than two inches of the point where the second seal is to be made. If the substance is a liquid it can more conveniently be introduced at a later stage.

Now bring the tube into the blowpipe flame gradually, and rotate it, while heating, at the place where it is to be closed. Allow the glass to soften and commence to run together until the diameter of the tube is reduced to about half its original size. Remove from the flame and draw the ends apart, this should give a long, thick extension as shown by f, Fig. 4. If any liquid is to be introduced, it may now be done by inserting a thin rubber or other tube through the opening and running the liquid in. A glass tube should be used with caution for introducing the liquid, as any hard substance will tend to scratch the inside of the glass and cause cracking. The final closure is made by melting the drawn-out extension in the blowpipe flame; the finished seal being shown by g, Fig. 4.

If the sealed tube has to stand internal

pressure, it is desirable to allow the glass to thicken somewhat more before drawing out, and the bottom seal should also be made thicker. For such a tube, and especially when it has to stand heating, as in a Carius determination of chlorine, each seal should be cooled very slowly by rotating it in a gas flame until the surface is covered with a thick layer of soot, and it should then be placed aside in a position where the hot glass will not come in contact with anything, and where it will be screened from all draughts.

Joining Tube.—We will now consider the various forms of join in glass tubing which are met with in the laboratory. First, as being easiest, we will deal with the end-to-end joining of two tubes of similar glass. *a*, *b*, and *c*, Fig. 5, illustrate this. One end of one of the tubes should be closed, a lip should be turned out on each of the ends to be joined, and both lips heated simultaneously until the glass is thoroughly soft. Now bring the lips together gently, until they are in contact at all points and there are no places at which air can escape; remove from the flame, and blow slowly and very cautiously until the joint is expanded as shown in *b*, Fig. 5. Reheat in the flame until the glass has run down to rather less than the original diameter of the tube, and give a final shaping by re-blowing. The chief

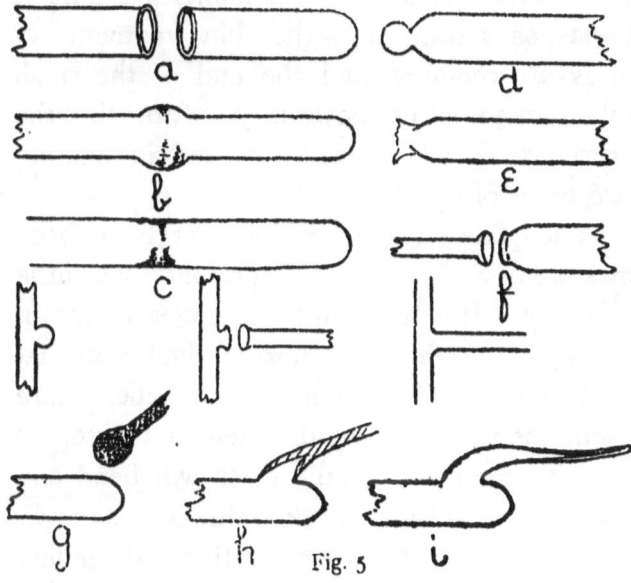

Fig. 5

factors of success in making such a join are, thorough heating of the glass before bringing the two tubes together, and avoidance of hard or sudden blowing when expanding the joint. The finished work is shown by *c*, Fig. 5.

To join a small glass tube to the end of a

large one, the large tube should first be sealed, a small spot on the extreme end of the seal heated, and air pressure used to expand the heated spot as shown in *d*. This expanded spot is then re-heated and blown out until it bursts as shown in *e*, the thin fragments of glass are removed and the end of the small tube turned out as shown in *f*. After this the procedure is similar to that used in jointing two tubes of equal size.

When these two forms of joint have been mastered, a piece will present but little difficulty. It is made in three stages as shown in Fig. 5, and the procedure is similar to that used in joining a large and small tube. Care should be taken to avoid softening the top of the "T" too much, or the glass will bend and distort the finished work; although a slight bend can be rectified by re-heating and bending back. Local re-heating is often useful in giving the joint its final shape.

An exhaustion branch is often made by a totally different method. This method is shown by *g*, *h*, and *i*, Fig. 5; *g* is the tube on which the branch is to be made. The end of a rod of similar glass should be heated until a mass of thoroughly liquid glass has collected, as shown, and at the same time a spot should be heated on that part of the tube where it is desired to make the branch. The mass of hot glass on the rod is now brought in contact with the heated spot on the tube and expanded by blowing as shown by h. The air pressure in the tube is still maintained while the rod is drawn away as shown by i. This will give a hollow branch which may be cut off at any desired point, and is then ready for connection to the vacuum pump.

If the rod used is of a dissimilar glass, the branch should be blown much thinner. Such a branch will often serve as a useful basis for joining two tubes of different composition, as the ordinary type of branch is more liable to crack when made with two glasses having different coefficients of expansion.

Blowing Bulbs.—A bulb may be blown on a closed tube such as that shown by c, Fig. 5, by rotating it in the blowpipe flame until the end is softened, removing it from the flame and blowing cautiously. It is desirable to continue the rotation during blowing. In the case of a very small tube, it is sufficient to melt the end without previous sealing, rotate it in the flame until enough glass has collected, remove from the flame and blow while keeping the tube in rotation.

Thermometer Bulbs.—If the thermometer is to be filled with mercury, it is desirable to use a rubber bulb for blowing, as moisture is liable to condense inside the tube when the mouth is used, and this moisture will cause the mercury thread to break. In any case, a slight pressure should be maintained inside the thermometer tube while it is in the flame; otherwise the fine capillary tube will close and it will be very difficult to expand the heated glass into a bulb.

Large Bulbs.—When a large bulb is needed on a small or medium sized tube, it is often necessary to provide more glass than would be obtained if the bulb were blown in the ordinary way. One method is to expand the tube in successive stages along its axis, as shown by *a*, Fig. 6. These expanded portions are then reheated, so that they run together into one hollow mass from which the bulb is blown; *b* and *c*, illustrate this. Another method, and one which is useful for very large bulbs, is to fuse on a length of large, thick-walled, tubing. The heat reflector, *g*, Fig. 3, should be used, if necesssary, when making large bulbs. It

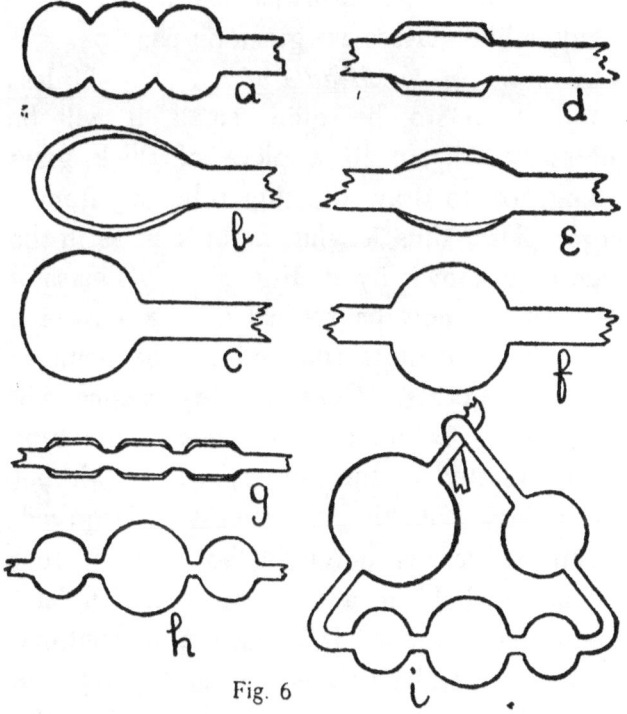

Fig. 6

consists of a sheet of asbestos mounted in a foot, and is used by being placed close to the mass of glass on the side away from the blow-pipe flame while the glass is being heated.

Bulbs of Dissimilar Glass.—These may be made by the second method given under "Large Bulbs," but the joint should be blown as thin as possible. Further instructions in the use of unlike glasses are given on page 94.

A Bulb in the Middle of a Tube.—Unless the bulb is to be quite small, it will be necessary to join in a piece of thick glass tubing, or to draw the thin tube out from a larger piece, thus leaving a thick mass in the middle as shown by *d*, Fig. 6. This mass of glass should now be rotated in the blowpipe flame until it is quite soft and on the point of running together. Considerable practice will be necessary before the two ends of the tube can be rotated at the same speed and without "wobbling," but this power must be acquired. When the glass is thoroughly hot, remove from the flame, hold in a horizontal position, and expand by blowing. It is essential to continue the rotation while this is done. Should one part of the bulb tend to expand more than the other, turn the expanded part to the bottom, pause for about a second, both in rotating and blowing, in order that the lower portion may be cooled by ascending air-currents; then continue blowing and turning as before.

Absorption Bulbs or Washing Bulbs.—These are made by an elaboration of the processes given in the last paragraph, *g*, *h*, and *i*, Fig. 6, illustrate this.

A Thistle Funnel.—This is made by blowing a fairly thick-walled bulb on a glass tube, bursting a hole by heating and blowing, and enlarging the burst-out part by heating and rotating against a turn-pin.

Bending Glass Tube.—Small tubing may be bent in a flat flame gas burner and offers no special difficulty. Large or thin-walled tubing should be heated in the blowpipe flame and a slight bend made; another zone of the tube, just touching the first bend, should now be heated and another slight bend made. In this way it is possible to avoid flattening and a bend having any required angle can gradually be produced. A final shaping of the bend may be made by heating in a large blowpipe flame and expanding slightly by air pressure.

Glass Spirals.—If a tube is heated by means of a long, flat-flame burner, the softened tube may be wound on to an iron mandrel which has previously been covered with asbestos. The mandrel should be made slightly conical in order to facilitate withdrawal. It is desirable to heat the surface of the asbestos almost to redness by means of a second burner, and thus avoid undue chilling

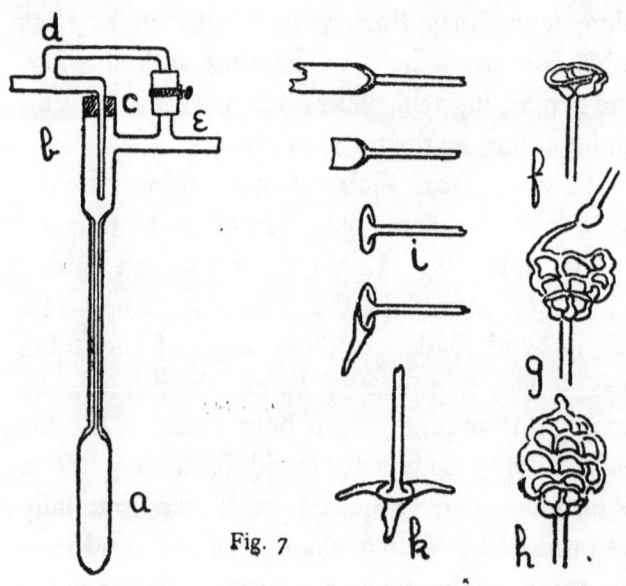

Fig. 7

of the glass and the consequent production of internal strain.

A Thermo-Regulator for Gas.—Fig. 7, *a*—*e*, shows an easily constructed thermo-regulator.

The mercury reservoir, *a*, and the upper part, *b*, are made by joining two larger pieces of tubing on to the capillary. The gas inlet passes through a rubber stopper, in order to allow of adjustment for depth of insertion, and the bye-pass branches, *d* and *e*, are connected by a piece of rubber tubing which can be compressed by means of a screw clip, thus providing a means of regulating the bye-pass.

Use of Glass Rod.—Apart from its most common laboratory use for stirring; glass rod may be used in building up such articles as insulating feet for electrical apparatus or acid-resisting cages for chemical purposes. Such a cage is shown by *f*, *g* and *h*, Fig. 7. Further, by an elaboration of the method of making an exhaustion branch, given on page 18, blown articles may also be constructed from rod. Note the added parts of *e*, Fig. 9.

A Simple Foot.—The form of foot shown by Fig. 7, *k*, is easy to make and has many uses. First join a glass rod to a length of glass tubing as shown (the joint should be expanded slightly by blowing), cut off the tube and heat the piece remaining on the rod until it can be turned out as shown by *i*. This should be done with the large turn-pin, and care should be taken not to heat the supporting rod too strongly, otherwise the piece of tube will become bent and distorted; it is better to commence by heating the edge of the piece of tube and turn out a lip, then extend the heating by degrees and turn out more and more until the foot looks like that shown by i.

We now need to make three projections of glass rod. These are produced as follows :—Heat the end of the glass rod until a thoroughly melted mass of glass has accumulated (the rod must be rotated while this is being done, otherwise the glass will drop off); when sufficient melted glass has been obtained, the edge of the turned-out foot should be heated to dull redness over about one-third of its circumference, and the melted glass on the rod should be drawn along the heated portion until both are so completely in contact as to form one mass of semi-fluid glass. The rod should now be drawn away slowly, and, finally, separated by melting off, thus producing a flat projection. A repetition of the process will give the other two projections, and the finished foot may be adjusted to stand upright by heating the

LABORATORY GLASS-BLOWING 27

projections slightly and standing it on the carbon plate mentioned on page 7. After the foot is adjusted it should be annealed slightly by heating to just below the softening point of the glass and then rotating in a smoky gas flame until it is covered with a deposit of carbon, after which it should be allowed to cool in a place free from draughts and where the hot glass will not come in contact with anything. The finished foot is shown by k, Fig. 7.

Building up from Glass Rod.—A glass skeleton-work can be constructed from rod without much difficulty, and is sometimes useful as a container for a substance which has to be treated with acid, or for similar purposes. The method is almost sufficiently explained by the illustration in Fig. 7; f shows the initial stage, g the method of construction of the net-work, and h the finished container. It is convenient to introduce the substance at the stage indicated by g. The important points to observe in making this contrivance are that the glass rod must be kept hot by working while it is actually in the flame, and that the skeleton must be made as thin as possible with the avoidance of heavy masses of glass at any place. If these details are neglected it will be almost certain to crack.

Stirrers.—These are usually made from glass rod, and no special instructions are necessary for their construction, except that the glass should be in a thoroughly fused condition before making any joins and the finished join should be annealed slightly by covering with a deposit of soot, as explained on page 16. The flat ends shown in *a*, Fig. 8, are made by squeezing the soft glass rod between two pieces of carbon, and should be reheated to dull redness after shaping. Fig. 8 also shows various forms of stirrer.

In order to carry out stirring operations in the presence of a gas or mixture of gases other than air, some form of gland or seal may be necessary where the stirrer passes through the bearing in which it runs. A flask to which is fitted a stirrer and gas seal is shown in section by *b*, Fig. 8. The liquid used in this seal may be mercury, petroleum, or any other that the experimental conditions indicate.

If the bearing for a stirrer is made of glass tube, it is desirable to lubricate rather freely;

otherwise heat will be produced by the friction of the stirrer and the tube will probably crack. Such lubrication may be supplied by turning out the top of the bearing tube and filling the

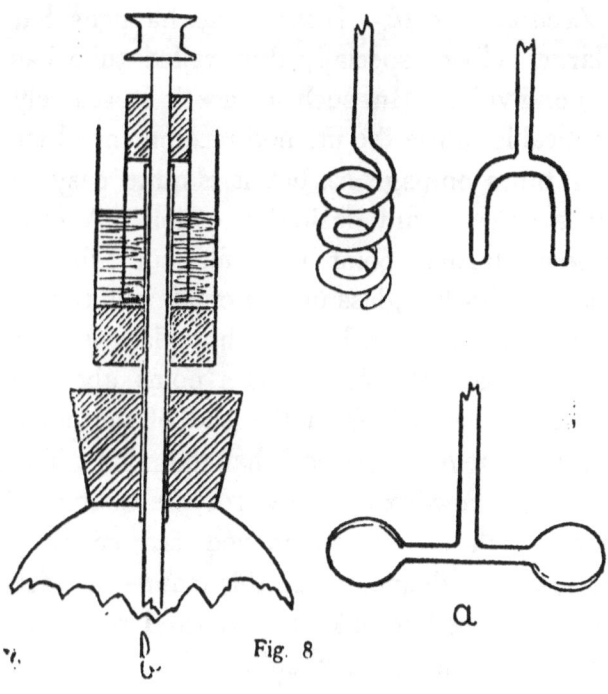

Fig. 8

turned-out portion with petroleum jelly mixed with a small quantity of finely ground or, better, colloidal graphite, and the bearing should also be lubricated with the same composition. Care should be taken not to employ so soft a lubricant or so large an excess as to cause it to run down the stirrer into the liquid which is being stirred.

Leading a Crack.—It sometimes happens that a large bulb or specially thin-walled tube has to be divided. In such a case it is scarcely practicable to use the method recommended for small tubes on page 12, but it is quite easy to lead a crack in any desired direction. A convenient starting point is a file cut; this is touched with hot glass until a crack is initiated. A small flame or a bead of hot glass is now used to heat the article at a point about a quarter of an inch from the end of the crack and in whatever direction it has to be led. The crack will now extend towards the source of heat, which should be moved farther away as the crack advances. In this manner a crack may be caused to take any desired path and can be led round a large bulb.

Cutting Glass with the Diamond.—Slips of window-glass can be used in place of glass rod for some purposes, and as cutting them involves the use of the glaziers' diamond or a wheel-cutter, they may well be mentioned under this heading.

In cutting a sheet of glass with the diamond, one needs a flat surface on which to rest the glass, and a rule against which to guide the diamond. The diamond should be held in an almost vertical position, and drawn over the surface of the glass with slight pressure. While this is being done the angle of the diamond should be changed by bringing the top of the handle forward until the sound changes from one of scratching to a clear singing note. When this happens the diamond is cutting. A few trials will teach the student the correct angle for the diamond with which he works, and the glass, if properly cut, will break easily. If the cut fails it is better to turn the glass over and make a corresponding cut on the other side rather than make any attempt to improve the original cut. The diamond is seldom used for cutting small glass tubes.

The use of the wheel-cutter calls for no special mention as it will cut at any angle, although the pressure required is somewhat greater than that needed by most diamonds.

CHAPTER III

Internal Seals, Air-Traps, Spray Arresters, Filter-Pumps—Sprays, Condensers; Plain, Double Surface, and Spherical—Soxhlet Tubes and Fat Extraction Apparatus—Vacuum Tubes, Electrode Work, Enclosed Thermometers, Alarm Thermometers, Recording Thermometers, "Spinning" Glass.

Internal Seals.—It is convenient to class those cases in which a glass tube passes *through* the wall of another tube or bulb under the heading of "Internal Seals." These are met with in barometers, spray arresters, and filter pumps, in condensers and some forms of vacuum tube. The two principal methods of making such seals will be considered first and their special application afterwards.

An Air Trap on a Barometer Tube.—This involves the use of the first method, and is perhaps the simplest example that can be given. Fig. 9, *a*, *a*1 and *a*2, show the stages by which this form of internal seal is made. For the first trials, it is well to work with fairly thick-walled

LABORATORY GLASS-BLOWING 33

tubing, which should be cut into two pieces, each being about eight inches long.

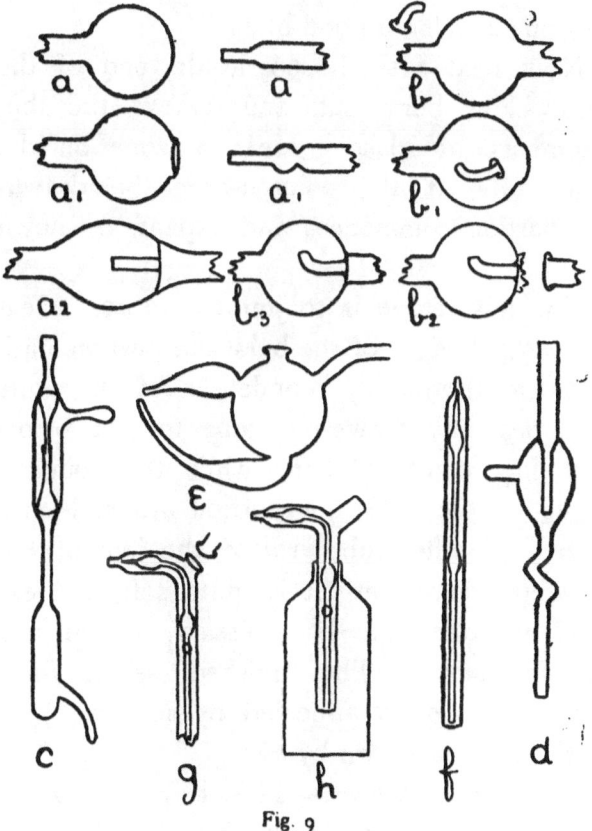

Fig. 9

First seal the end of one tube as described on page 13, heat the sealed end and expand to a thick walled bulb. Fuse the end of the other tube, attach a piece of glass rod to serve as a handle, and draw out; cut off the drawn-out portion: leaving an end like *a*.

Now heat a small spot at the end of the bulb, blow, burst out, and remove the thin fragments of glass. Heat a zone on the other tube at the point where the drawn-out portion commences and expand as shown by *a*1.

The next stage is to join the tubes. Heat the ragged edges of the burst-out portion until they are thoroughly rounded. At the same time heat the drawn-out tube to just below softening point. Then, while the rounded edges of the burst-out portion are still soft, insert the other tube; rotate the join in the blowpipe flame until it is quite soft, and expand by blowing. If necessary, reheat and expand again. The finished seal, which should be slightly annealed by smoking in a sooty flame, is shown by *a*2.

A Spray Arrester.—This is made by the second method, in which the piece of tube which projects inside the bulb is fused in position first and the outer tube is then joined on. The various stages of making are illustrated by b, b_1 and b_2, Fig. 9.

A bulb is blown between two tubes by the method given on page 22, the larger tube is then cut off and the small piece of tube introduced into the bulb after having been shaped as shown in by b, Fig. 9. The opening in the bulb is sealed as shown by b_1. The sealed part is now heated and the bulb inclined downwards until the inner tube comes in contact with the seal and is fused in position. This operation requires some practice in order to prevent the inner tube either falling through the soft glass or becoming unsymmetrical. The end of the bulb, where the inner tube comes in contact with it, is now perforated by heating and blowing, thus giving the form shown by b_2, and the outer tube is joined on. The finished spray arrester is shown by b_3. Practice alone will give the power to produce a symmetrical and stable piece of work.

Two Forms of Filter Pump.—That illustrated by d, Fig. 9, is made by the method explained under "An Air Trap on a Barometer Tube." That illustrated by c is made by the method explained under "A Spray

Arrester." No new manipulation is involved, and the construction should be clear from a study of the drawings.

Multiple and Branched Internal Seals.—A fuller consideration of these will be found on page 39, but one general principle may well be borne in mind; that, as far as is possible, a tube having both ends fastened inside another tube or bulb should be curved or have a spiral or bulb at some point in its length, otherwise any expansion or contraction will put great strain on the joints.

Sprays.—A spray which is easy to make, easy to adjust, and easy to clean after use is shown by *e*, Fig. 9. The opening on the top of the bulb is made by melting on a bead of glass, expanding, bursting, and fusing the ragged edges. The two branches which form the spray producing junction are made by the method used for an exhaustion branch and described on page 18.

A spray which can be introduced through the neck of a bottle is shown by *h*, Fig. 9. The various stages in making this are illustrated by *f*, and *g*. If the inner tube is made by drawing out from a larger piece of glass so that two supporting pieces are left on each side of the place where it is intended to make the final bend, that bend can be made in a flat-flame gas burner without causing the inner tube to come in contact with the walls of the outer tube. Care must be taken when joining on the side piece that the inner tube is not heated enough to fuse it. The small hole in the side of the outer tube is produced by heating and bursting.

A Liebig's Condenser.—This consists of a straight glass tube passing through an outer cooling jacket. In practice it is better to make the jacket as a separate piece, and to effect a water-tight junction by means of two short rubber tubes. It may, however, be made with two internal seals of the class described under "A Spray Arrester." There is much less risk of these seals cracking if the inner tube is made in the form of a spiral or has a number of bulbs blown on it in order to give a certain amount of elasticity.

A Double-Surface Condenser.—Fig. 10 shows a condenser of this nature which is supplied by Messrs. Baird and Tatlock. It may be built up in stages as shown by *a, b,*

38   LABORATORY GLASS-BLOWING

and *c*, but the work involved requires considerable skill, and the majority of laboratory workers will find it cheaper to buy than to make.

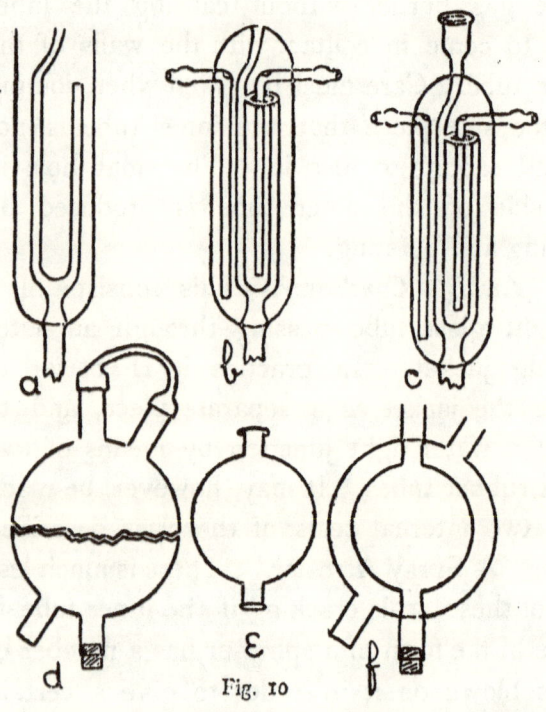

Fig. 10

A Spherical Condenser.—Such a condenser as that shown by *f*, Fig 10, involves a method which may find application in a number of cases. The outer bulb is blown from a thick piece of tubing which has been inserted in a smaller piece (see *d*, Fig. 6); then the inner bulb by similar method. It is now necessary to introduce the smaller bulb into the larger, and for this purpose the larger bulb must be cut into halves. A small but deep cut is made with the file or glass-blowers' knife in the middle of the larger bulb, and at right angles to the axis of the tube on which it is blown. A minute bead of intensely heated glass is now brought in contact with the cut in order to start a crack. This crack may now be led round the bulb as described on page 30. If the work is carried out with care, it is possible to obtain the bulb in two halves as shown by *d*, and these two halves will correspond so exactly that when the cut edges are placed in contact they will be almost air-tight. The two tubes from the smaller bulb should be cut to such a length that they will just rest inside the larger, and the ends should be expanded. Place the inner bulb in position and fit the two halves of the outer bulb together, taking great care not to chip the edges. If the length of the tubes on the inner bulb has been adjusted properly, the inner bulb will be supported in position by their contact with the tubes on the outer bulb.

Now rotate the cracked portion of the outer bulb in front of a blowpipe flame and press the halves together very gently as the glass softens. Expand slightly by blowing if necessary. If a small pin-hole develops at the joint it is sometimes possible to close this with a bead of hot glass; but if the bulb has been cut properly there should be no pin-holes formed. The condenser is finished by joining on the side tubes and sealing the inner tube through by the methods already given. In order to blow bulbs large enough to make a useful condenser, it will be convenient to employ the multiple-jet blowpipe described on page 4.

A Soxhlet-Tube or Extraction Apparatus.—This involves the construction of a re-entrant join where the syphon flows into the lower tube. It is of considerable value as an exercise and the complete apparatus is easy to make.

A large tube is sealed at the bottom and the top is lipped, as in making a test-tube. A smaller tube is then joined on by a method similar to that given on page 18, but without making a perforation in the bottom of the large tube. Heating and expanding by air pressure, first through the large tube, then through the

LABORATORY GLASS-BLOWING 41

smaller tube and then again through the large tube, will give a satisfactory finish to this part of the work.

The syphon tube is now joined on to the large tube as shown by *a*, Fig. 11, care being

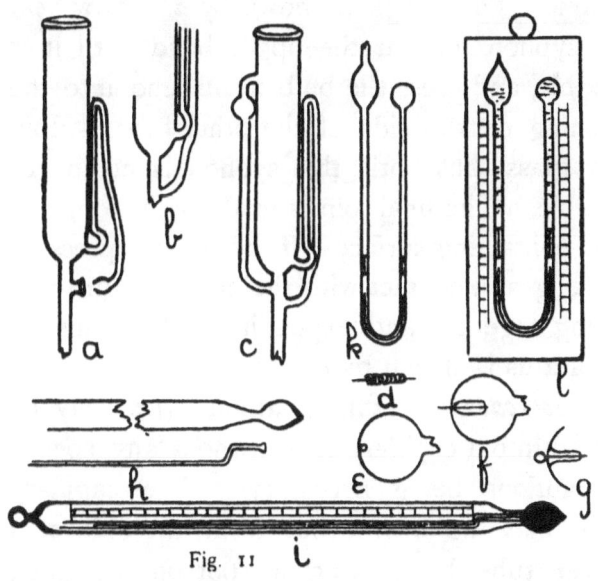

Fig. 11

taken to seal the other end of the syphon tube before joining. The details of the final and re-entrant joint of the syphon tube are shown at the lower part of *a*. This join is made by expanding the sealed end of the syphon tube into a small, thick-walled bulb, and the bottom of this bulb is burst out by local heating and blowing; the fragments of glass are removed and the edges made smooth by melting. A similar operation is carried out on the side of the tube to which the syphon tube is to be joined. This stage is shown by *a*. Now heat the syphon tube at the upper bend until it is flexible, and press the bulb at its end into the opening on the side of the other tube. Hold the glass thus until the syphon is no longer flexible. The final join is made by heating the two contacting surfaces, if necessary pressing the edges in contact with the end of a turn-pin, fusing together and expanding. The finished apparatus is shown by *c*.

Electrodes.—A thin platinum wire may be sealed into a capillary tube without any special precautions being necessary. The capillary tube may be drawn out from the side of a larger tube by heating a spot on the glass, touching with a glass rod and drawing the rod away; or the exhaustion branch described on page 18 may be used for the introduction of an electrode. It is convenient sometimes to carry out the exhaustion through the same tube that will afterwards serve for the electrode. The electrode wire is laid inside the branch before connecting to the exhaustion pump. When exhaustion is completed the tube is heated until the soft glass flows round the platinum and makes the seal air-tight. The branch is now cut off close to the seal on the pump side, a loop is made in the projecting end of the platinum wire, and the seal is finished by melting the cut-off end.

Platinum is usually employed for such work, but if care is taken to avoid oxidation it is not impossible to make fairly satisfactory seals with clean iron or nickel wire. Hard rods of fine graphite, such as are used in some pencils, may also be sealed into glass, but it seems probable that air would diffuse through the graphite in the course of time.

Another method for the introduction of an electrode is illustrated by d, e, f and g, Fig. 11. In this case the bulb or thin-walled tube into which the electrode is to be sealed is perforated by a quick stab with an intensely heated wire—preferably of platinum—which is then withdrawn before the glass has had time to harden, and thus a minute circular hole is made. The electrode is coated with a layer of similar glass, or of the specially made enamel which is sold for this purpose, inserted into the bulb or tube by any convenient opening, and adjusted by careful shaking until the platinum wire projects through the small hole. The bulb or tube is then fused to the coating of the electrode and the whole spot expanded slightly by blowing. The appearance of the finished seal is shown by *g*. It is well to anneal slightly by smoking.

Thermometers.—Apart from the notes on page 20 with respect to the blowing of a suitable bulb on capillary tubing there is little to say in connection with the glass working needed in making a plain thermometer. The size desirable for the bulb will be determined by the bore of the capillary tube, the coefficient of expansion of the liquid used for filling, and the range of temperature for which the thermometer is intended.

Filling may be carried out as follows:—Fit a small funnel to the open end of the capillary by means of a rubber tube, and pour into the funnel rather more than enough of the liquid to be used than is required to fill the bulb. Mercury or alcohol will be used in practice, most probably. Warm the bulb until a few air bubbles have escaped through the liquid and then allow to cool. This will suck a certain amount of liquid into the bulb. Now heat the bulb again, and at the same time heat the capillary tube over a second burner. The liquid will boil and sweep out the residual air, but it is necessary to heat the capillary tube as well in order to prevent condensation. Allow the bulb and tube to cool, then repeat the heating once more. By this time the bulb and tube should be free from air, and cooling should give a completely filled thermometer. Remove the funnel and heat the thermometer to a few degrees above the maximum temperature for which it is to be used; the mercury or other filling liquid will overflow from the top, and, as the temperature falls, will recede, thus allowing the end of the capillary to be drawn out. Reheat again until the liquid rises to the top of the tube, then seal by means of the blowpipe flame. The thermometer is now finished except for graduation; this is dealt with on page 75.

An Alarm Thermometer.—A thermometer which will complete an electric circuit when a certain temperature is reached may be made by sealing an electrode in the bulb and introducing a wire into the top, which in this case is not sealed. Naturally, this thermometer will be filled with mercury. There is considerable difficulty in filling such a bulb without causing it to crack.

Several elaborations of this form are made, in which electrodes are sealed through the walls of the capillary tube, thus making it possible to detect electrically the variation of temperature when it exceeds any given limits.

An Enclosed or Floating Thermometer.— The construction of this type of thermometer is shown by h and i, Fig 11. It is made in the following stages :—A bulb is blown on the drawn-out end of a thin-walled tube as shown by h. A small bulb is blown on the end of a capillary tube, burst, and turned out to form a lip which will rest in the drawn-out part of the thin-walled tube but is just too large to enter the bulb. The capillary tube is introduced and sealed in position, care being taken to expand the joint a little. The thermometer is filled and the top of the capillary tube closed by the use of a small blowpipe flame. A paper scale having the necessary graduations is inserted, and the top of the outer tube is closed as shown by i.

A Maximum and Minimum Thermometer.—
If a small dumb-bell-shaped rod of glass or metal is introduced into the capillary tube of a horizontally placed, mercury-filled thermometer in such a position that the rising mercury column will come in contact with it, the rod will be pushed forward. When the mercury falls again the rod will be left behind and thus indicate the maximum temperature attained. If a similar dumb-bell-shaped rod is introduced into an alcohol-filled thermometer and pushed down until it is within the alcohol column, it will be drawn down by surface tension as the column falls; but the rising column will flow passed it without causing any displacement; thus the minimum temperature will be recorded.

Six's combined maximum and minimum thermometer is shown by b, Fig. 11. In this case both maximum and minimum records are obtained from a mercury column, although the thermometer bulb is filled with alcohol. It is an advantage to make the dumb-bell-shaped rods of iron, as the thermometer can then be reset by the use of a small magnet, another advantage consequent on the use of metal being that the rods can be easily adjusted, by slight bending, so as to remain stationary in the tubes when the thermometer is hanging vertically, and yet to move with sufficient freedom to yield to the pressure of the recording column.

The thermometer may be filled by the following method :—When the straight tube has been made the first dumb-bell is introduced and shaken down well towards the lower bulb, the tube is now bent to its final shape and the whole thermometer filled with alcohol as described on page 44. Now heat the thermometer to a little above the maximum temperature that it is intended to record, and pour clean mercury into the open bulb while holding the thermometer vertically. Allow to cool, and the mercury will be sucked down. The second dumb-bell is now introduced, sufficient alcohol being allowed to remain in the open bulb to about half fill it, and the alcohol in this bulb is boiled to expel air. The tube through which the bulb was filled in now sealed.

Clinical Thermometers.—The clinical thermometer is a maximum thermometer of a different type. In this case there is a constriction of the bore at a point just above the bulb. When the mercury in the bulb commences to contract, the mercury column breaks at the constriction and remains stationary in the tube, thus showing the maximum temperature to which it has risen.

Vacuum Tubes.—There are so many forms of these that it is scarcely practicable or desirable to give detailed instructions for making them; but an application of the various methods of glass-working which have already been explained should enable the student to construct most of the simpler varieties. An interesting vacuum tube is made which has no electrodes, but contains a quantity of mercury. When the tube is rocked so as to cause friction between the mercury and the glass sufficient charge is produced to cause the tube to glow.

A Sprengel Pump.—This, in its simplest form, is illustrated by *a*, Fig. 12. Such a form, although highly satisfactory in action, needs constant watching while in action, as should the mercury funnel become empty air will enter the exhausted vessel. Obviously, the fall-tube must be made not less than thirty inches long; the measurement being taken from the junction of the exhaustion branch with the fall-tube to the top of the turned-up end.

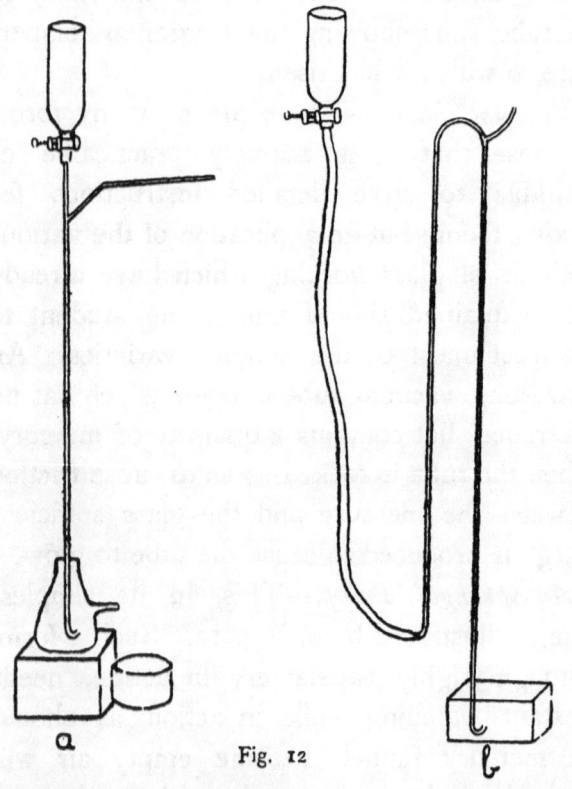

Fig. 12

A Macleod Pump.—One form of this is illustrated by *b*, Fig. 12. It has the advantage that the mercury reservoir may be allowed to become empty without affecting the vacuum in the vessel being exhausted.

"*Spinning*" *Glass.*—By the use of suitable appliances, it is quite possible to draw out a continuous thread of glass, which is so thin as to have almost the flexibility and apparent softness of woollen fibre; a mass of such threads constitutes the "glass wool" of commerce.

The appliances necessary are:—a blowpipe capable of giving a well-formed flame of about six or eight inches in length, a wheel of from eighteen inches to three feet in diameter and having a flat rim of about three inches wide, and a device for rotating the wheel at a speed of about three hundred revolutions per minute.

A very satisfactory arrangement may be made from an old bicycle; the back wheel having the tyre removed and a flat rim of tin fastened on in its place. The chain drive should be retained, but one of the cranks removed and a handle substituted for the remaining pedal. The whole device is shown by Fig. 13.

The procedure in "spinning" glass is as

follows:—First melt the end of a glass rod and obtain a large mass of thoroughly softened glass, now spin the wheel at such a speed that its own momemtum will keep it spinning for several seconds. Touch the end of the melted rod with another piece of glass and, without

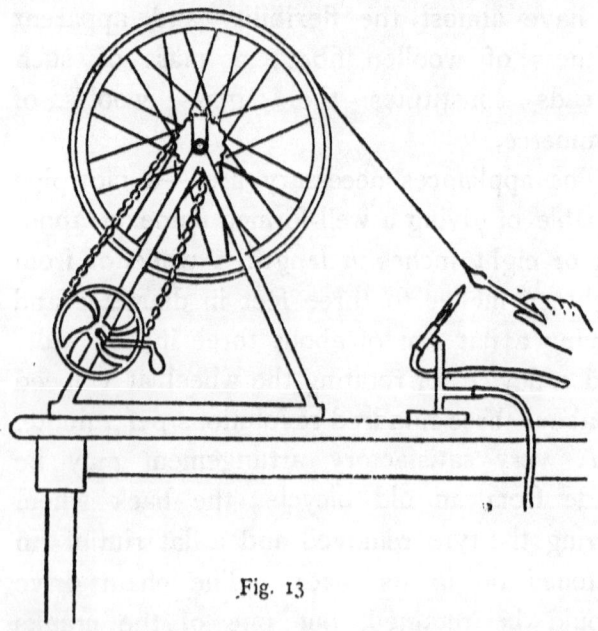

Fig. 13

withdrawing the original rod from the blowpipe flame, draw out a thread of molten glass and twist it round the spinning wheel. If this is done properly, the thread of glass will grip on the flat rim, and by continuing to turn the wheel by hand it is possible to draw out a continuous thread from the melted rod, which must be advanced in the blowpipe flame as it is drawn on the wheel. If the rod is not advanced sufficiently the thread will melt off, if it is advanced too much, so as to heat the thick part and allow the glass to become too cool at the point of drawing out, then the thread will become too thick, but it is easy after a little practice to obtain the right conditions. Practice is necessary also in order to find the right speed for the wheel.

When sufficient glass has been "spun," the whole "hank" of thin thread may be removed by drawing the thumb-nail across the wheel at any point on its flat rim, thus breaking the threads, and allowing the "hank" to open.

Brushes for Use with Strong Acids.—Glass wool, if of fine enough texture to be highly flexible, can be used to make acid-resisting brushes. A convenient method for mounting the spun glass is to melt the ends of the threads together into a bead, and then to fuse the bead on to a rod; thus giving a brush. If a pointed brush is necessary, the point may be ground on an ordinary grindstone or carborundum wheel by pressing the loose end of the spun glass against the grinding wheel with a thin piece of cardboard.

When using brushes of this description, it is well to bear in mind the fact that there is always a liability of a few threads of glass breaking off during use.

CHAPTER IV

Glass, Its Composition and Characteristics. Annealing. Drilling, Grinding, and Shaping Glass by methods other than Fusion. Stopcocks. Marking Glass. Calibration and Graduation of Apparatus. Thermometers. Exhaustion of Apparatus. Joining Glass and Metal. Silvering Glass.

THERE are three kinds of glass rod and tubing which are easily obtainable; these are soda-glass, which is that usually supplied by chemical apparatus dealers when no particular glass is specified; combustion-glass, which is supplied for work requiring a glass that does not so easily soften or fuse as soda-glass; and lead-glass, which is less common. There are also resistance-glass, made for use where very slight solubility in water or other solutions is desirable, and a number of other special glasses; but of these soda-glass, combustion-glass, lead-glass, and resistance-glass are the most important to the glass-blower whose work is connected with laboratory needs.

Soda-Glass.—Consists chiefly of sodium

silicate, but contains smaller quantities of aluminum silicate, and often of calcium silicate; there may also be traces of several other compounds.

The ordinary soda-glass tubing melts easily in the blowpipe flame, it has not a long intermediate or viscous stage during fusion, but becomes highly fluid rather suddenly; it does not blacken in the reducing flame. Bad soda-glass or that which has been kept for many years, tends to devitrify when worked. That is to say the glass becomes more or less crystalline and infusible while it is in the flame; and in this case it is often impossible to do good work with that particular sample of glass; although the devitrification may sometimes be remedied by heating the devitrified glass to a higher temperature. The presence of aluminum compounds appears to have some influence on the tendency of the glass to resist devitrification. Soda-glass, as a rule, is more liable to crack by sudden heating than lead-glass, and articles made from soda-glass often tend to crack spontaneouly if badly made or, in the case of heavier and thicker articles, if insufficiently annealed.

Combustion-Glass.—Is usually a glass containing more calcium silicate and potassium silicate than the ordinary "soft" soda-glass. It is much less fusible than ordinary soda-glass, and passes through a longer intermediate or viscous stage when heated. Such a glass is not very suitable for use with the blowpipe owing to the difficulty experienced in obtaining a sufficiently high temperature. If, however, a certain amount of oxygen is mixed with the air used in producing the blowpipe flame this difficulty is minimised.

Resistance-Glass.—May contain zinc, magnesium, and other substances. As a rule it is harder than ordinary soda-glass, and less suitable for working in the blowpipe flame. It should have very little tendency to dissolve in water, and hence is used when traces of alkali or silicates would prove injurious in the solutions for which the glass vessels are to be used.

Lead-Glass.—This, or "flint" glass as it is often called from the fact that silica in the form of crushed and calcined flint was often used in making the English lead-glasses, contains a considerable proportion of lead silicate. Such a glass has, usually, a particularly bright appearance, a high refractive index, and is specially suitable for the production of the heavy "cut-glass" ware.

Lead-glass tubing is easy to work in the blowpipe flame, melts easily, but does not become fluid quite so suddenly as most soda-glasses; articles made from it are remarkably stable and free from tendency to spontaneous cracking, although, as is essential for all the heavy or "glass-house" work, the massive articles need annealing in the oven.

The two chief disadvantages of lead-glass for laboratory work are that it is blackened by the reducing gases if held too near to the blue cone of the blowpipe flame, and that it is rather easily attacked by chemical reagents; thus ammonium sulphide will cause blackening.

The effect of the reducing flame on lead is not altogether a disadvantage, however; because a little care in adjusting the blowpipe and a little care in holding the glass in the right position will enable the student to work lead-glass without producing the faintest trace of blackening. This, in addition to being a valuable exercise in manipulation, will teach him to keep his blowpipe in good order, and prove a useful aid in his early efforts to judge as to the condition of the flame. It prevents discouragement if the student does his preliminary work with the soda-glass, but he should certainly make experiments with lead-glass as soon as he has acquired reasonable dexterity with soda-glass.

Annealing.—Annealing is a process by which any condition of strain which has been set up in a glass article, either by rapid cooling of one part while another part still remains hot, or by the application of mechanical stress after cooling is relieved. Annealing is carried out by subjecting the article to a temperature just below the softening point of the glass, maintaining that temperature until the whole article has become heated through the thicker part, and then reducing the temperature very gradually; thus avoiding any marked cooling of the thinner and outer parts first.

For thin glass apparatus of the lamp-blown or blowpipe-made variety in which there are no marked difference of thickness, such as joins on tubes, ordinary seals, bulbs, etc., there is little need for annealing; and even those having rather marked changes of thickness, such as filter pumps, can be annealed sufficiently by taking care that the last step in making is heating to just below visible redness in the blowpipe flame and then rotating in a sooty gas flame until covered with a deposit of carbon. The article should then be allowed to cool in a place free from draughts and where the hot glass will not come in contact with anything.

A few of the blowpipe-made articles, such, for example, as glass stopcocks, need more careful annealing, and for this purpose a small sheet-iron oven which can be heated to dull redness over a collection of gas burners will serve. Better still, a small clay muffle can be used. In either case, the article to be annealed should be laid on a clean, smooth, fireclay surface, the temperature should be maintained at a very dull red for two or three hours and then reduced steadily until the oven is cold. This cooling should take anything from three to twelve hours, according to the nature of the article to be annealed. A thick article, or one having great irregularities in thickness will need much longer annealing than one thinner

LABORATORY GLASS-BLOWING 61

or more regular. As a rule, soda-glass will need more annealing than lead-glass.

Drilling Glass.—Small holes may be drilled in glass by means of a rod of hard steel which has been broken off, thus giving a more or less irregular and crystalline end.

There are several conditions necessary to enable the drilling of small holes to be carried out successfully :—the first of these is that the "drill" should be driven at a high speed. This may be done by means of a geared hand-drill such as the American pattern drill, although a somewhat higher speed than this will give is even more satisfactory. The second condition is that the pressure on the drill is neither too light nor too heavy; this is conveniently regulated by hand. The third condition is that the drill be prevented from "straying" over the surface of the glass; for this purpose a small metal guide is useful. The fourth condition is that a suitable lubricant be used; a strong solution of camphor in oil of turpentine is perhaps the most suitable. For commercial work, a diamond drill is often used, but this is scarcely necessary for the occasional work of a laboratory.

Larger Holes in Glass.—The method of drilling with a hard steel rod is not highly satisfactory for anything but small holes. When a larger hole, say one of an eighth of an inch or more, is needed it is better to use a copper or brass tube. This tube may be held in an American hand-drill, but a mixture of carborundum or emery and water is supplied to the rotating end. Tube or drill must be lifted at frequent intervals in order to allow a fresh supply of the grinding material to reach the end. In this case, also, a guide is quite essential in the early stages of drilling; otherwise the end of the tube will stray. The speed of cutting may be increased slightly by making a number of radial slots in the end of the tube; these serve to hold a supply of the grinding material.

Grinding Lenses.—This is scarcely within the scope of a book on glass-blowing for laboratory purposes, but it may be said that the lens may be ground by means of a permutating mould of hard lead or type-metal. The rough shaping is done with coarse carborundum or emery, and successive stages are carried on with finer and finer material. The last polishing is by the use of jewellers' rouge on the mould, now lined with a fine textile.

Filing Glass.—If a new file, thoroughly lubricated with a solution of camphor in oil of turpentine, is used, there is but little difficulty in filing the softer glasses. A slow movement of the file, without excessive pressure but without allowing the file to slip, is desirable. After a time the cutting edges of the file teeth will wear down and it will be necessary to replace the file by another.

Grinding Stoppers.—This is, perhaps, the most common form of grinding that the laboratory worker will need to perform, and for that reason, rather full details of the proceedure are desirable.

A very crude form of ground-in stopper may be made by drawing out the neck and the mass of glass which is intended to form the stopper to approximately corresponding angles, wetting the surfaces with a mixture of the abrasive material and water, and grinding the stopper in by hand. Frequent lifting of the stopper is necessary during grinding, in order to allow fresh supplies of abrasive material to reach the contacts. When an approximate fit is obtained, the coarse abrasive should be washed off, care being taken that the washing is complete, and a finer abrasive substituted. After a while, this is replaced in its turn by a still finer grinding material.

Such a method of grinding may give a satisfactory stoppering if the angles of the plug and socket correspond very closely before grinding is commenced; but if there is a wide difference in the original angles, then no amount of grinding by this method will produce a good result. The reason for this is that the plug will become so worn in the preliminary grinding as to assume the form of a highly truncated cone; the socket will assume a reverse form, and the end result will be a loose-fitting plug and socket.

Satisfactory grinding may be carried out by the use of copper or type-metal cones for the preliminary shaping. Such cones should be mounted on a mandrel which will fit into the chuck of the American hand-drill and turned on the lathe to the desirable angle for stoppering. A number of these cones will be necessary. A number of similar moulds, that is to say blocks of type-metal or hard lead in which is a hole corresponding in size and angle

LABORATORY GLASS-BLOWING 65

to the plug desired, should be made also. These must be rotated, either in the lathe or by other means, and are used for the preliminary shaping of the plug. If but few plugs are to be ground it is unnecessary to provide a means of rotating the moulds, as the plug may be held in the hand and ground into the mould in a manner similar to that used in the first method of stoppering.

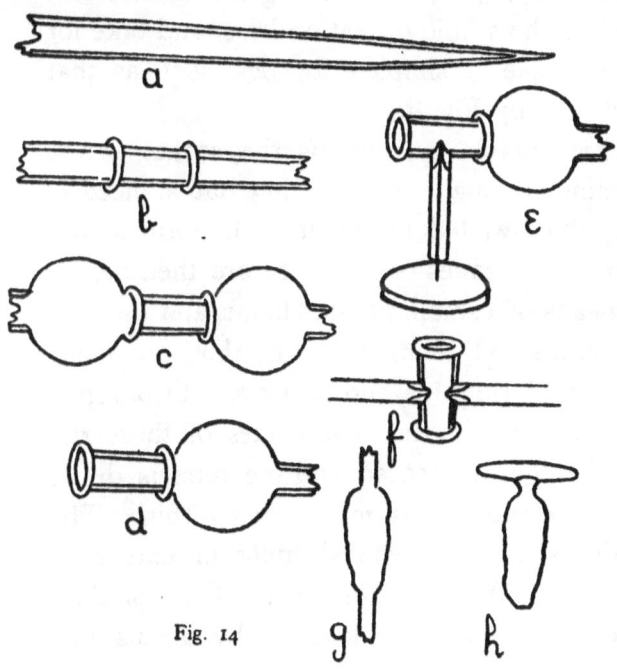

Fig. 14

When the socket and plug have been ground, by the successive use of cones and moulds, to the desired angle, so that they correspond almost exactly, the plug is given its final fitting into the socket by grinding-in with a fine abrasive, in the manner first described.

Stopcocks.—Although it would be more strictly in keeping with the form of this book to divide the making of stopcocks into two parts; shaping by heat and grinding, we will consider the whole operation here, and take for our example a simple stopcock such as that illustrated by Fig. 14.

The "blank," *f*, that is the socket before grinding, is made by drawing out a piece of fairly thick-walled tubing into the form shown by *a*. Two zones on this tube are then heated by means of a small, pointed flame, and the tube is compressed along its axis, thus producing two raised rings as shown by *b*. Two zones, slightly towards the outer sides of these two raised rings are heated and the tube is drawn while air pressure is maintained within. This produces two thin-walled bulbs or extensions similar to those shown by *c*. One of these extensions is now broken off by means of a

LABORATORY GLASS-BLOWING

sharp blow with the edge of a file or other piece of metal, and the edges of the broken glass are rounded in the flame. The other extension is left to serve as a handle. We have now a piece of glass like that shown by *d*. Now heat a spot on the side of this, medially between the raised rings, until the glass is on the point of becoming deformed, and bring the intensely heated end of a smaller tube in contact with the heated spot. Without disturbing the relative positions of the two tubes, press the smaller tube down on a thin steel wire, so that the wire passes along the tube and enters the soft glass; thus forming a projection inside the sockets as shown by *e*. The wire must be withdrawn, again immediately. When the wire has been withdrawn, heat the place where it entered to dull redness, in order to relieve any strain; break off the thin extension, which up to the present has served as a handle, round off the broken edges in the flame, and join on and indent a similar piece of small tubing to the opposite side of the socket; the socket at this stage being shown by *f*. The "blank" for the socket is now completed, but it must be heated to dull redness in order to relieve strain and be placed in an annealing oven, where it should be annealed for some hours.

The "blank" for the plug offers no special difficulty; it is made by heating a glass rod and compressing it axially until a mass having the form shown by g, Fig. 14, is produced; the end of this is heated intensely and brought in contact with the rather less heated side of a glass tube which has been drawn to the shape desired for the handle; when contact is made a slight air pressure is maintained in the glass tube, thus producing a hollow join. The ends of the tube are sealed and the bottom of the plug is drawn off, thus giving the finished "blank" as shown by h. This blank is now held in a pair of asbestos-covered tongs, heated to dull redness all over, and transferred to the annealing oven.

When cold, the socket is ground out by the second method given under "Grinding Stoppers"; that is to say, by means of type-metal or copper cone, and the plug is ground to fit in a corresponding mould. When the fit is almost perfect, the transverse hole is drilled in the plug, and the final finishing is made with fine abrasive powder. Great care must be taken in the final grinding that there is no accumulation of abrasive material in the transverse hole of the plug ; if this is allowed to occur there will be a ring ground out of the socket where the holes move, and the tightness of the finished stopcock will be lost.

Marking Glass.—As a preliminary to a consideration of the methods of graduating and calibrating glass apparatus, it is convenient to consider the various methods which are available for marking glass. Among these are, the writing diamond, the carborundum or abrasive pencil, the cutting-wheel, and etching by means of hydrofluoric acid. Each produces a different class of marking and each is worthy of independent consideration.

The Writing Diamond.—This is the name given to a small irregular fragment of " bort " which is usually mounted in a thin brass rod. Such a diamond, if properly selected, has none of the characteristics of a cutting diamond ; although one occasionally finds so-called " writing diamonds " which will produce a definite cut. These should be rejected.

The writing diamond is used in much the same way as a pencil, but is held more perpendicularly to the object, and a certain amount of pressure is necessary. The mark produced is a thin scratch which, although fairly definite, lacks breadth, and this is a disadvantage where the marking has to be read at a distance. This disadvantage may to some extent be overcome by making a number of parrallel scratches.

The Abrasive Pencil.—A rod of carborundum composition may be ground or filed to a point, and this forms a very useful pencil for general work. The marking produced is rather less definite than that produced by a writing diamond, but has the advantage of being broader.

The Cutting Wheel.—" Cutting " in this case is scarcely the ideal expression, it should rather be "grinding," but "cutting" is more commonly used. Exceedingly good graduations may be made by the edge of a small, thin, abrasive wheel which is mounted on the end of a small mandrel and driven by a flexible shaft from an electric motor or any other convenient source of power. The depth of the mark can be controlled, and very light pressure will suffice.

Etching.—This is often the quickest and

easiest way of marking glass apparatus. The object to be marked should first be warmed and coated very thoroughly with a thin film of paraffin wax. When cold, the marking is made through the paraffin wax by means of a needle point, and the object is then exposed to the action of hydrofluoric acid. If a shallow but clearly visible marking is desired, it is well to use the vapour of the acid; this may be done by bending up a sheet-lead trough on which the object can rest with the marked surface downwards. A little of the commercial hydrofluoric acid, or a mixture of a fluoride and sulphuric acid, is distributed over the bottom of the trough, and the whole arrangement is allowed to stand for about an hour. The object is washed thoroughly and the paraffin wax removed, either by melting and wiping off or by the use of a solvent, and the marking is finished.

If a deep marking is desired, in order that it may afterwards be filled with some pigment, a better result is obtained by the use of liquid commercial hydrofluoric acid, which is a solution of hydrogen fluoride in water. The acid is mopped on to the object after the markings have been made on the paraffin wax film, and allowed to remain in contact for a few minutes. It is advantageous to repeat the mopping-on process at intervals during the etching.

In all cases where hydrofluoric acid is used, or stored, it is of great importance to keep it well away from any optical instruments, as the most minute trace of vapour in the air will produce a highly destructive corrosion of any glass surfaces.

Methods of Calibration.—In the case of apparatus for volumetric work, this is usually carried out by weighing, although some of the smaller subdivisions are often made by measurement. When the subdivisions are made in this way it is of importance to see that the walls of the tube or vessel to be calibrated are parallel. Great errors arise in some of the commercial apparatus from neglect of this precaution. A convenient method of testing for parallelism, in the case of a wide tube, is to close one end and to weigh in successive quantities of mercury. An observation of the length occupied by each successive quantity will indicate any change in the bore. In the case of capillary tubes, *it is* convenient to introduce an unweighed quantity of mercury, measure its length accurately, and then to move it along the tube in stages, either by tilting the tube or by the application of air pressure. A measurement of the length at each stage will indicate whether the bore is approximately parallel or not. Neither of these methods is to be relied on without a careful examination of the tube, as it may happen that there are local irregularities in the bore which compensate for each other, and do not, therefore, affect the volume of a given length. Obviously, the smaller the quantity of mercury with which the test is carried out and the greater the number of observations made, the less risk will there be of such an error. A liquid, such as water or alcohol, which wets the glass is not suitable for such a test, unless special precautions are taken.

When, however, a pipette or burette has to be calibrated to deliver a certain volume of water, the final calibration must be made with this liquid. Thus, the burette would first be calibrated by weighing in definite quantities of mercury of say 13·54 grammes (1 cc at $15^{\circ C}$.), each of the 1 cc divisions should be marked by some temporary marking. The burette is now filled with a solution of potassium bichromate and sulphuric acid and allowed to soak for some time; the bichromate is washed out and distilled water is put in. Successive quantities of water are run out of the jet, a fixed time being allowed for draining, and the weights of the quantities delivered are noted. This procedure will give the necessary data for altering the marking so that it may correspond to 1 cc *delivered*. Each 1 cc division is now divided into tenths by the method described below. A final verification of the markings should be made when the subdivision is completed.

Subdivision of Graduations.—Mark out the spaces to be subdivided on a sheet of paper. Take a reliable ruler on which any convenient length is divided into the desired number and place it across the lines at such an angle that the limits noted on the rule exactly bridge the gap. Now draw parallel lines through the markings.

Copying a Scale.—When a scale has been prepared on paper and it is necessary to copy that scale on the waxed-glass surface for etching, a convenient method is to employ a

LABORATORY GLASS-BLOWING 75

long wooden bar having a sharp needle passing through it at either end. The scale and object to be marked are fastened in line with one another, and the caliper bar is used from step to step. The mark is made by moving the bar through a minute portion of a circle, which provided that the bar is two or three feet in length, will not introduce any perceptible error in a scale of say a quarter of an inch in width. The arrangement is shown by Fig. 15.

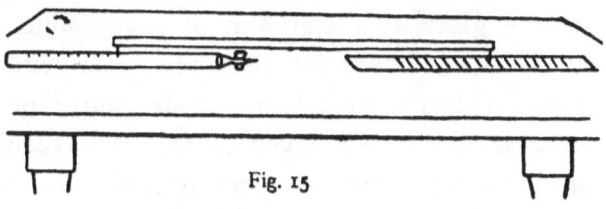

Fig. 15

Graduating a Thermometer.—Assuming that the thermometer has been made of carefully selected tubing in which the bore is parallel and free from any small irregularities, we have only to fix the freezing point and boiling point. The intervening space may then be divided into 100 (if the thermometer is to be Centigrade) or 180 (if Farenheit). This division may be carried out by the method given under

"Subdivisions of Graduations." A thermometer should not be calibrated until some weeks after making, as the glass bulb tends to contract.

Joining Glass and Metal.—It sometimes happens that one needs to make a more permanent and less flexible joint between a glass and metal tube than can be obtained by means of a rubber tube. To this end, any one of three slightly different methods may be employed. In the method of Chatelier one first coats the glass with platinum or silver, which may be done by moistening the glass with platinum chloride or silver nitrate and then heating to redness; a layer of copper is then deposited electrolytically on the treated surface of the glass, and soldering is carried out in the usual manner.

McKelvy and Taylor call attention to two other methods in the *Journal of the Chemical Society* for September, 1920. In one of these methods the glass is coated with platinum by covering it with a suspension of platinum chloride in oil of lavender and heating until the oil is burnt off. The metal tube is then tinned on its inner side and soldered to the prepared glass, slightly acid zinc chloride being used as a flux.

In the second method, a joint is made by means of the Kraus flux, which consists of equal weights of zinc oxide, borax, and powdered soda-glass fused together. This is coated on the inner surface of the metal tube, and the hot glass tube, which has had the end slightly flanged to give support, is inserted. Fusion of the flux is completed by heating the outside of metal tube.

Silvering Glass.—In all cases where it is intended to deposit a silver mirror on a glass surface, thorough cleaning is essential. Prolonged soaking in a hot solution of potassium bichromate which has been acidified with sulphuric acid will often prove useful. The glass should then be washed thoroughly, rinsed in distilled water, and the solution should then be used.

There are many formulæ for the silvering solution, but that used in Martin's method may be given :—

A—Nitrate of Silver	40 grammes
Distilled Water	1000 c. cm.
B—Nitrate of Ammonium	60 grammes
Distilled Water	1000 c. cm.
C—Pure Caustic Potash	100 grammes
Distilled Water	1000 c. cm.
D—Pure Sugar Candy	100 grammes
Distilled Water	1000 c. cm.

Dissolve and add :—

Tartaric Acid	23 grammes

Boil for ten minutes, and when cool add :—

Alcohol	200 c. cm.
Distilled Water to	2000 c. cm.

For use take equal parts of A and B. Mix together also equal parts of C and D in another vessel. Then mix both liquids together in the silvering vessel and suspend the glass to be silvered face downwards in the solution. Or if a vessel has to be silvered on the inside, the solution is poured in. In this case, the deposition of silver may be hastened by immersing the vessel to be silvered in warm water.

In working with a silver solution containing ammonia or ammonium salts there is sometimes the possibility of forming an explosive silver compound. It is well, therefore, to avoid keeping such solutions longer than is necessary, and to bear in mind that any deposit formed by solutions containing both silver and ammonia may have explosive properties, especially when dry.

CHAPTER V

Extemporised Glass-Blowing Apparatus—The Use of Oil or other Fuels—Making Small Rods and Tubes from Glass Scrap—The Examination of Manufactured Apparatus with the View to Discovering the Methods Used in Manufacture—Summary of Conditions Necessary for Successful Glass-Blowing.

IF, in the early stages of his study of glass-blowing, the student should attempt to work with the very simplest appliances, it is probable that his progress will be hindered; the use of the apparatus will require an undue amount of care and his attention will be distracted from the actual manipulation of the glass. The case is widely different after he has acquired a certain facility in glass-blowing.

A Simple Form of Blowpipe.—Although there are even more simple forms than that described here, we are not concerned with them. The form described is the simplest with which any considerable amount of glass-blowing can be carried out with certainty.

This form consists of a tube through which air may be blown with the mouth, a condensation chamber in which any moisture from the breath can condense, a blowpipe jet, a supporting piece and a source of flame.

The tube, condensation chamber, and jet are combined in the ordinary Black's blowpipe, such as is used for blowpipe tests in qualitative analysis; it consists of a conical tin tube having a mouthpiece at the small end and a side tube which carries a brass jet. A support for such a blowpipe may be cut out of a piece of brass or tin-plate, and should be fastened to a small, flat, wooden board. A source of flame may consist of an ordinary brass elbow, such as is used on gas fittings, and into which a piece of thin brass tube (the body of a fish-tail burner from which the perforated non-metallic plug has been removed will serve quite well) has been fitted. It is an advantage to flatten the brass tube somewhat and to file the flattened end to a slope which corresponds with the angle at which the blowpipe jet enters the burner. The whole source of the flame should be mounted on a separate base, in order that it may be moved while adjusting the apparatus to the best relative positions of flame and blowpipe jet. The complete apparatus is shown by *a*, Fig. 16.

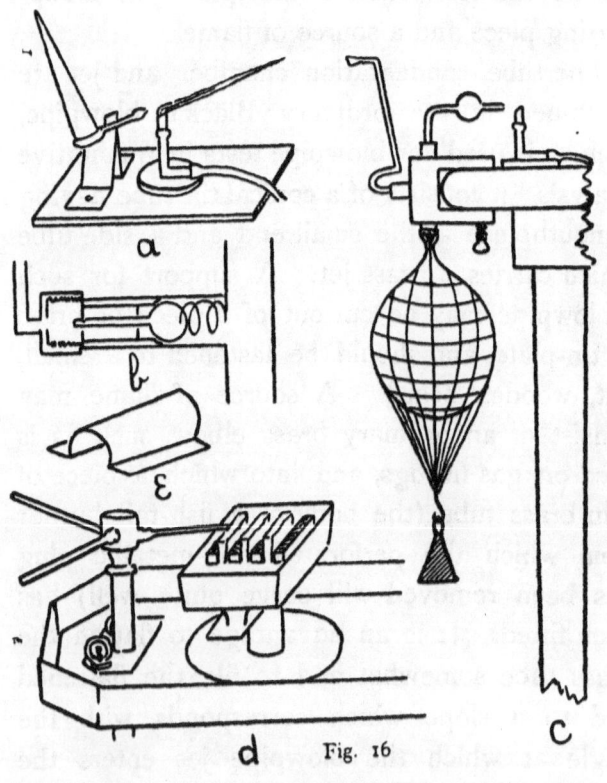

Fig. 16

In order to take full advantage of this blowpipe, it is desirable that the student should learn to maintain a steady steam of air with his

LABORATORY GLASS-BLOWING 83

mouth and, at the same time, be able to breathe. This requires a little practice.

As a first exercise in breathing, before trying to breathe while using the mouth blowpipe, the student should close his mouth and inflate his cheeks with air; now, still keeping his cheeks tightly inflated, he should attempt to breathe through the nose. At first, this may be found rather difficult, but it becomes remarkably easy after a little practice. When he has mastered this, the student may practise the same operation, but with the blowpipe. It is important to bear in mind that the cheeks, not the lungs, form the reservoir for air used in maintaining the blowpipe flame. After a while, the student will find that he can maintain a steady air pressure and yet breathe with complete comfort.

In adjusting the flame, care should be taken not to blow so hard as to produce a ragged and noisy cone of fire. A small jet, such as that commonly used on a mouth blowpipe, will with care give a pointed and quiet flame, having an appearance similar to that shown in the illustration.

With a blowpipe like this, it is quite easy to seal glass tubes up to an inch in diameter, to join tubes up to half an inch in diameter, to bend tubes, to blow small bulbs, and to make the simpler forms of internal seal; but the provision for condensation of moisture is not ideal, and prolonged use of such a blowpipe also tends to produce undue fatigue.

A Mouth Blowpipe With an Expanding Reservoir.—This form of blowpipe can be made to give most excellent results; it is highly portable, and does not produce nearly so much fatigue when used continuously as the blowpipe described in the last section. Various slight modifications have been made in its construction during the last eighty years, but that described below will be found quite satisfactory.

The apparatus consists of a tube through which air is blown from the mouth, a valve through which the air passes into an expanding reservoir, and a blowpipe jet in communication with the reservoir.

In making the valve, several essentials have to be remembered; it must allow a free passage of air into the reservoir, it must open easily, and must close quickly. A satisfactory form of valve is that shown by b, Fig 16. The moving part consists of a light glass bulb of about three-eights of an inch diameter and having a glass stem of rather under one-eighth diameter and about an inch and a half long. This stem rests in a guide at the end of a brass tube, the bulb contacting against the other end which is approximately shaped. The bulb and its seating are ground air-tight. A very light spring holds the bulb in position.

This valve is fitted into a metal or glass T piece, one limb of which leads to the air reservoir and the other limb leads to the blow-pipe jet; the limb containing the valve leads to the tube through which the air is blown in.

A convenient reservoir may be made from a fairly large football bladder. A network of string should be fitted over the outside of the bladder and the strings should terminate in a hook on which a weight can be hung, in order to provide a means of adjusting the pressure at which the air is delivered to the jet. This bladder should be washed out and allowed to drain after use.

The air tube which passes from the valve to the mouth may conveniently be made of brass, but, in order to avoid the continued contact of metal with the lips of the operator, it should be fitted with a non-metallic mouthpiece. It is an advantage from the point of view of portability to have the air tube easily detachable from the T piece containing the valve.

The blowpipe jets, of which there may be several with advantage, may be made of glass tubing, bent to the most convenient angle and having an enlargement or bulb at some point in the tube. This bulb serves as a final condensing place for any traces of moisture that may escape from the larger reservoir.

The whole device, blowing tube, reservoir, and T piece may be fastened to a clamp, so that it can be secured on the edge of any table where blowpipe work is to be carried out. If the blowpipe is to be used with gas, the form of burner described under "A Simple Form of Blowpipe" will be found quite satisfactory.

The Use of Oil, or Other Non-Gaseous Fuels.—Although gas, when available, is usually preferred on account of its convenience, there are several other fuels which give a hotter flame. They have, also, the additional advantage of not requiring any connecting pipes; but each has its own disadvantage.

One liquid fuel deserves special mention as

being rather less desirable than the others ; this is alcohol. Although very convenient in use, it has the disadvantage of being rather too highly inflammable and capable of burning without a wick, thus involving a certain fire risk; the flame is scarcely visible in a bright light, and the heat given by either the ordinary flame or the blowpipe flame produced from alcohol is considerably less than that from a similar flame in which coal gas is used, For small work, however, the facility with which a spirit lamp may be lighted may more than counterbalance these disadvantages at times.

Paraffin Wax.—Where there is no coal gas available and the blowpipe is only required at intervals, and especially where high portability is required, there are few fuels so convenient as paraffin wax. This may be obtained in pieces of a satisfactory size by cutting paraffin candles, from which the wick has been withdrawn, into lengths of about half an inch. These cut pieces have the advantage over any oily fuel, such as colza oil, that they can be wrapped in paper or carried in a cardboard box ; further they will keep indefinitely, even in the presence of air, without undergoing any perceptible change.

Forms of Lamp for Paraffin Wax.—Probably, the best form is that devised by Thomas Bolas, and described by him in the *Journal of the Society of Arts*, December 2nd, 1898. This lamp consists of a small open tray of iron, through which pass three or more flat tubes, and between these tubes are placed small flat pieces of wick, the fit being such that the pieces of wick may be adjusted easily by means of a pair of pointed tweezers.

The flame thus obtained, instead of having one large hollow, is broken or divided so that the combustion is concentrated into a smaller area, and the air blast, which is directed across the flame, carries the flame with it in a more complete manner than is the case with the ordinary flame; a more thorough combustion being realised by this arrangement.

Another advantage is the ease with which the wick may be changed and a larger or smaller wick inserted to suit the flame to any size of air jet.

This form of lamp may be used for oily fuel, although it is specially suitable for paraffin wax.

Two small pieces of bent tin-plate may be used as side covers, and these serve to adjust the flame within certain limits. A tin-plate cover which fits easily over the whole lamp serves as an extinguisher. The complete lamp is shown by d, Fig. 16, and this figure shows also a quick-change air-jet device, the whole arrangement forming a blowpipe for use where a non-gaseous fuel is to be employed.

Although the lamp just described is desirable when complete control over the size of the flame is necessary, and if the ideal conditions and maximum heat are to be obtained, yet a simpler form of lamp will be found to give very good results. Such a lamp may consist of a flat tin tray, having a diameter of about three and a half inches and a depth of about one inch. In this tray is a tin support for the wick, and the wick itself may consist of a bundle of soft cotton, for example, a loosely rolled piece of cotton cloth, but in either case the top of the wick should be cut to approximately the same angle as that at which the blowpipe jet meets the flame.

In using paraffin wax as a fuel, it is necessary to see that sufficient wax reaches the wick to prevent charring during the first few minutes before the bulk of the wax is melted.

Animal and Vegetable Oils.—Almost any oil may be used as a fuel, but many tend to become hard and gummy if allowed to stand in the air for any considerable time. When this happens, the wick becomes clogged and it is impossible to obtain a good flame. A number of the oils tend, also, to produce rather strongly smelling smoke.

A Flame-Guard for Use With Non-Gaseous Fuels.—In order to avoid the eye-strain produced by the luminous base of the flame from a wick burning paraffin wax or oil, it is often advantageous to make a small tunnel of tinplate, which can be rested on the sides of the lamp and rises over the top of the wick. Such a flame guard is shown by e, Fig 16.

Small Rods and Tubes from Glass Scrap:—It is scarcely practicable to make small quantities of good glass with the blowpipe flame as the only source of heat, but it is less difficult to make small rods or tubes from glass scrap, and the ability to do this is sometimes of considerable value when a small tube has to be joined on to some special piece of apparatus made of glass of unknown composition. It may be possible to obtain some fragments of similar glass, either from a broken part of the apparatus or from a similar piece, and from these fragments small tubes or rods can be made.

The fragments of glass may be melted together on the end of a clay pipe-stem, care being taken to avoid trapping air bubbles as fresh fragments are added to the molten mass. When a sufficient quantity of glass has been accumulated, the viscous mass may be drawn out into a rod by bringing another pipe-stem into contact with the hot mass, rotating both pipe-stems steadily, and separating them until a rod of the desired size has been obtained.

If, on the other hand, it is desired to produce a tube from the mass of heated glass, the mass should be blown hollow before the pipe-stems supporting it are separated.

Methods of Manufacture.—When the student has familiarised himself with the more common operations and processes used in glass-blowing, he will be in a position to increase his skill and knowledge of special methods by a critical examination of various examples of commercial work. There are few exercises more valuable than such an examination, combined with an attempt to reconstruct the stages and the methods by which the article chosen for examination was made.

Obviously, it is impossible to give full details of all constructions in a small text-book; but it is easy to give an example of the constructional methods employed in the making of almost any piece of light blown-glass apparatus, and these methods should prove of special value when apparatus of a new pattern has to be evolved for the purposes of research. That is to say, one designs the apparatus required, applies known methods of construction as far as possible, and, by the examination of commercial apparatus having similar features, evolves the new methods required. For an exercise in such a process of reconstruction we may well take an ordinary commercial vacuum tube, such as that shown by *a*, Fig. 17.

In the tube from which this drawing was made, it was found that the spiral in the middle bulb was of a slightly yellowish colour and gave a green fluorescence when the electric discharge was passed through the tube; that is to say, the spiral is made of uranium-glass, which is usually a soda-glass containing trace of uranium, and hence differing slightly in composition from the ordinary glasses. The

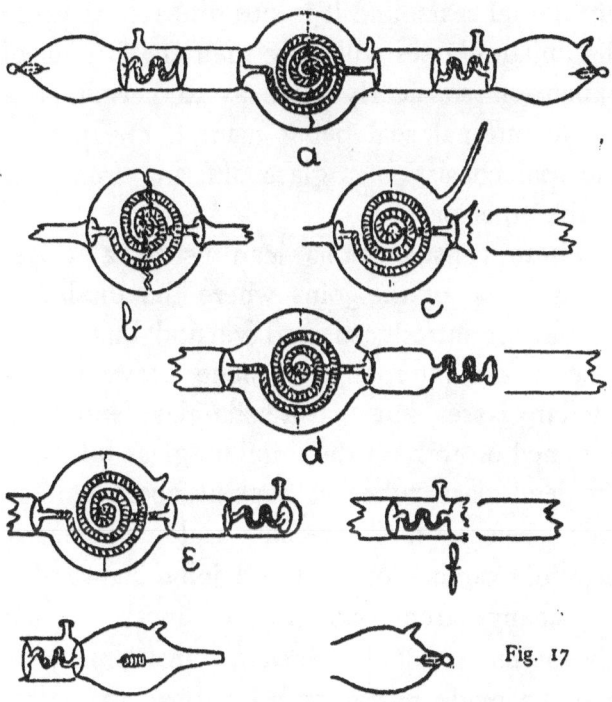

Fig. 17

two enclosed tubes which are bent into a series of S bends gave a pink fluorescence, which indicates lead-glass; and the remainder of the tube fluoresced with an apple-green colour;

this suggests ordinary soda-glass. We have, therefore, a piece of apparatus in which three dissimilar glasses are joined, while, at the same time, that apparatus contains a number of internal seals, and it is not probable that the dissimilar glasses will have their coefficients of expansion so nearly alike as to permit of a stable internal seal being made if one part of the seal consists of a glass differing from that of the other part.

These considerations lead us to a closer examination of the joins where the dissimilar glasses are introduced, and we find that in no case is the internal seal made between dissimilar glasses, but that a soda-glass extension is joined on to both the uranium-glass tube and the lead-glass tubes at a point about half an inch before the internal seal commences. Careful examination of these joins shows that the change from one glass to another is not abrupt but gradual. Such a transitional joint may be made by taking a length of soda-glass tubing, sealing the end and fusing a minute bead of the other glass on to the sealed end, the end is then expanded and another bead of the other glass added, this bead is expanded

LABORATORY GLASS-BLOWING 95

and the operation is repeated, thus building up a tube, and, finally, the tube of the other glass is joined on to the end of this.

We are now concerned with the question of the insertion of the uranium-glass spiral into the bulb (see p. 38). Obviously the spiral is too large to pass through the necks of the bulb, and it is difficult to imagine that the spiral was obtained by the insertion of a length of straight tubing which was bent after entering the bulb; therefore, the only remaining method is that the spiral was made first and the soda-glass extensions fastened on, and that the bulb was blown, cut in halves and the spiral inserted, and the two halves were then rejoined. That this was actually the case is confirmed by traces of a join which are just visible round the middle of the bulb. The insertion of the spiral and the making of the first internal seal are shown by *b*, and *c*.

There is one detail in making the second join of the spiral to the bulb which calls for attention, and the small branch, similar to an exhaustion branch, at the side of the bulb provides a clue to this. If an attempt were made to complete the second internal seal through a closed bulb it would be impossible to obtain a good result, as the air-pressure in the bulb would not be under control when once union was effected, and further heating of the air in the bulb would cause expansion and perforate the wall near the second internal seal; we therefore make a small branch which can be left open and through which such air-pressure as may be found necessary can be maintained.

The third join, by which the lead-glass tube is joined to the soda-glass is made in stages similar to those in which the soda-glass and uranium-glass were joined; but the internal seal is most conveniently made by sliding a length of tubing over the lead-glass and fusing this tubing to the large diameter soda-glass tube to which the lead-glass is already joined. The first stage of this operation is illustrated by *d*. When this seal is completed, the end of the soda-glass tube is drawn off and sealed as shown in *e*, and at this stage a side tube or branch is joined on. The sealed end of the outer and large diameter soda-glass tube is heated until it contracts and fuses to the enlargement that has previously been joined to the lead-glass tube, and the end is burst out as shown in *f*. Another length of soda-glass is then joined on to the burst-out end, and this length of soda-glass tubing is drawn out to a thin-walled contraction; the non-contracted part is expanded to form the bulb, and a small exhaustion branch made on the side, the drawn-out portion being cut off, and an electrode, previously prepared by coating a part of its length with a suitable enamel, is introduced. The tube is tilted to keep the electrode away from the drawn-out end, which is melted off and sealed. A small perforation is made with a hot platinum or iron wire in the sealed end, the electrode is shaken into position, and the sealing is completed as explained on page 42.

The remainder of the tube, that is to say the lead-glass tube and the bulb on the other side of the middle bulb, is completed in a similar manner.

SUMMARY OF CONDITIONS NECESSARY FOR SUCCESS IN GLASS-BLOWING.

For the convenience of the student, it may be well to summarise the chief essentials for success in glass-blowing, and at the same time to add such brief notes on the various methods as may seem desirable.

Adjustment of Blowpipe.—The air jet should be clean internally, and so centered as to give a flame having a well-defined blue portion, the tip of the flame should not be only slightly luminous but purple in colour. In the case of a blowpipe burning oil or wax fuel the flame may be a trifle more ragged without disadvantage.

Bellows and Blowing.—The bellows should be adjusted to deliver air at constant pressure, either by insertion of a tap or, better, by attention to the wind reservoir if necessary. The movement of the foot in blowing should be steady, not jerky.

Heating Glass.—The tube or rod should be heated cautiously until it has reached its softening point in its thickest part. Steady rotation of the glass during the heating is almost essential.

Blowing a Bulb or Expanding a Join.—Prolonged heating is necessary in order that the thick parts may be heated completely through. Blowing should take place by stages, in order that the thin parts, which tend to expand first, have time to cool. The thick parts can then be expanded by further blowing and thus a bulb or expansion of even thickness can be obtained.

Cutting Glass.—The most useful method for general use is by means of the file or glass-blowers' knife. Either file or knife must be kept sharp by grinding. Neither file nor knife should be used on hot glass. The diamond and wheel cutter are useful for cutting sheet-glass, and when the diamond is employed a singing noise is an indication of a satisfactory cut.

Leading a Crack.—A crack may be led in any desired direction by means of a bead of hot glass or a small gas flame. The glass which it is desired to crack should be heated at a point slightly in advance of the crack, which will extend in the direction of the source of the heat.

Turning Out the End of a Tube.—This is done by heating the end of the tube and rotating it against an iron rod. The rod must be kept polished and free from rust, and it must not be allowed to become too hot while in use, otherwise the glass will stick to it.

Joining Unlike Glasses.—Joints between unlike glasses are often unstable. When such joints are made it is desirable to blow them as thin as possible, and to avoid the junction of unlike glasses in any complex joint, such as an internal seal. A transitional portion of tubing may be built up by the successive addition and interfusion of beads of one of the glasses to the end of a sealed tube consisting of the other glass.

Joining a Tube to a Very Thin Bulb.—The bulb may be thickened at the point of union by fusing on a bead of glass and expanding this slightly. A small central portion of the expanded part may then be perforated by bursting and the tube joined on.

Insertion of One Bulb Within Another.—A bulb may be divided into two halves by leading a crack round it and the inner bulb is then introduced. The two halves of the outer bulb may be fitted together (care being taken to avoid any damage to the edges), and the bulb may be completed by rotating the contacting edges before the blowpipe until they are soft, and then expanding slightly by means of air-pressure.

Annealing.—For most purposes, in the case of thin, blowpipe-made or lamp-blown glass apparatus, it is sufficient to cool slowly by rotating the finished article over a smoky flame and setting it aside in a place free from draughts, and where the hot glass will not come in contact with anything.

Simple bulbs and joints do not even need this smoking; but thick articles, and especially those that are to be subjected to the stress of grinding, need more prolonged annealing in a special oven.

Use of Lead-Glass.—When lead-glass is to be used, the blowpipe flame should be in good adjustment and the glass should not be allowed to approach so near to the blue cone as to be blackened. Slight blackening may often be removed by heating the glass in the extreme end of the flame.

Lead-glass articles tend to be rather more stable than similar articles of soda-glass.

Combustion-Glass.—This may be worked more easily if a small percentage of oxygen is introduced into the air with which the blowpipe flame is produced. If the air is replaced entirely by oxygen there is a risk of damaging the blowpipe jet, unless a special blowpipe is employed.

Internal Seal.—There are two ways of making these, one, in which the inner portion of the tube is fused on to the inside of the bulb or tube through which it is to pass, an opening is made by bursting and the outer tube is joined on. This is a quick and in some ways more satisfactory method than the other, in which there is no separate inner piece.

Rubber Blowing Tube.—In complicated work it is often convenient to use a thin rubber blowing-tube which is connected with the work either by a cork and piece of glass tubing or by fitting over a drawn-out end. The use of such a blowing-tube avoids the inconvenience of raising the work to the mouth when internal air-pressure is required. One end of the rubber tube is retained in the mouth during work.

General Notes.—A large amount of glass-blowing is spoiled through carelessness in arranging the work beforehand. The student should have every detail of his manipulation clearly in mind before he commences the work;

he should not trust to evolving the method during the actual manipulation.

Undue haste is another fruitful source of failure. Practically every operation in glass-blowing can be carried out in a perfectly leisurely manner, and it is better to err rather on the side of deliberation than on the side of haste.

If, as will doubtless happen at times, a piece of work gives trouble and it is necessary to pause and consider the whole question, or if for any other reason it is necessary to stop during the construction of a partially finished join or other operation, great care should be taken not to allow the work to cool. A large, brush-like flame may be produced by increasing the amount of gas admitted to the blowpipe, and the work should be held just in front of the current of hot air produced by such a flame.

It will then be possible to continue work on this without causing it to crack when further heat is applied.

As time goes on, the student will find an increasing confidence in his ability to manipulate the soft glass, and with increasing

104 LABORATORY GLASS-BLOWING

confidence will come rapidly increasing power of manipulation. Perhaps the greatest obstacle to success in glass-blowing is undue haste in manipulation.

INDEX

Absorption bulbs, 21, 23.
Airtube, flexible, 8, 102.
Alarm thermometer, 45.
Annealing, 7, 60.

Bellows, adjusting pressure of, 5, 6.
Bellows, foot, 5, 6.
Bending tubes, 23.
Blackening, 58, 101.
Branching, 18, 19.
Brushes of spun glass, 53.
Blowpipe flame, quality of, 3.
Blowpipe for mouth blast, 80, 82, 84.
Blowpipe, for paraffin wax, 82, 88.
Blowpipe, Herepath's, 2.
Blowpipe jet, centring, 3, 98.
Blowpipe jet, dirt in, 3.
Blowpipe jet, multiple, 4, 40.
Blowpipe, Letcher's, change, 4.
Blowpipe, simple form of, 80.
Bulb, medially on tube, 22.
Bulbs, 19, 20, 22, 38, 98.
Bulbs, absorption, (Liebig's), 21, 23.
Bulbs, dividing, 39, 95.
Bulbs from rod, 25.
Bulbs, internal, 38.
Bulbs, thick, 21.
Cages, from glass rod, 24, 25, 27.
Calibration, 72.
Carius tubes, 16.
Condenser, Liebig's, 37.
Condensers, various, 37, 38.

Cone, carbon, 8.
Crack, leading, 30, 99.
Cracking, subversive, 103.
Cutting glass with diamond, 30.
Cutting tubes, 11, 99.

Diamond (glazier's), use of, 30.
Dissimilar glass, joining of, 22, 94.
Drilling, 61.

Electrodes, sealing in, 42, 97.
Etching glass, 70.
Extemporised appliances, 80.
Examination of apparatus, 93.

Failure, Haste chief Source of, 103.
Failures, Notes as to, 97.
File, with oblique ground edge, 7.
Filing glass, 63.
Filter pumps, 35
Foot, 25
Fuels various, 82, 86, 87, 89.
Funnel, thistle, 23.
General principles and precautions, 1, 97
Glass, varieties of, 9, 55, 91-97.
Graduation, 72-76

Haste, Source of Failure, 103
Heat reflector, asbestos, 7.
Heating, intensive, 7, 57.
Heating precautions, 12, 98

Joining dissimilar glass, 22,

INDEX

Joining glass to metal, 76
Joining tubes, 16, 94, 100.

Knife, Glass blower's, 7, 99.

Lenses, grinding, 63.

Marking glass, 69
Methods, analytic study of, 91, 93.

Oxygen for intensive heating, 57, 101.

Precautions and General Principles, 1, 97.
Pumps, Filter, 35.
Pumps, Sprengel, 49, 50.

Re-entering branch, 40.
Reflector of heat, asbestos, 7.
Rod, uses and articles from, 17, 25, 27, 28.
Rod, blowing to hollow, 17, 25, 26, 91.

Scrap glass, working, 90.

Sealing tubes, 12, 13, 14.
Sealed tubes for pressure, 15, 16.
Sealing in of Electrodes, 42, 97.
Seals, internal (airtraps), 32, 102.
Silvering glass, 77.
Soldering glass, 76.
Soxhlet-tube, 40.
Spirals, 23, 95.
Spray arrester, 34.
Spray producers, 36.
Sprengel pumps, 49, 50.
Spinning glass, 51.
Stopcocks, 60, 66.
Stoppering, 63.
Stirrers, 28, 29.
Summary as to precautions and failures, 97.

Taps, 60, 66.
Thermometers, Various, 44-49.
Thermo-regulator, 24.
Thistle Funnel, 23.
Tools, Various small, 7.
Turn-pins, 7, 8, 99.
Turning out open ends, 14, 99.

Printed in Great Britain by
W. Jolly & Sons, Ltd., Printers, Aberdeen.

www.ingramcontent.com/pod-product-compliance
Lightning Source LLC
Chambersburg PA
CBHW020753230426
43665CB00009B/574